FOR WALLY TRABING
with gallons of good wishes

Jim and Charlie

APRIL 1980

OTHER BOOKS BY JAMES D. HOUSTON

Fiction:

BETWEEN BATTLES (Dial Press, 1968)
GIG (Dial Press, 1969)
A NATIVE SON OF THE GOLDEN WEST (Dial Press, 1971)
AN OCCURRENCE AT NORMAN'S BURGER CASTLE (Yes! Capra Chapbook, 1972)
CONTINENTAL DRIFT (Knopf, 1978)

Non-fiction:

FAREWELL TO MANZANAR, with Jeanne Wakatsuki Houston (San Francisco Book Co./Houghton Mifflin, 1973)
WRITING FROM THE INSIDE (Addison-Wesley Co., 1973)
OPEN FIELD, with John R. Brodie (San Francisco Book Co./Houghton Mifflin, 1974)
THREE SONGS FOR MY FATHER (Yes! Capra Chapbook, 1974)

Collections Edited:

CALIFORNIA HEARTLAND, WRITING FROM THE GREAT CENTRAL VALLEY, with Gerald Haslam (Capra Press, 1978)
WEST COAST FICTION: MODERN WRITING FROM CALIFORNIA, OREGON, AND WASHINGTON (Bantam Books, 1979)

Drawing by: Philip Garner

Gasoline

The Automotive Adventures of Charlie Bates

James D. Houston

A Noel Young Book

CAPRA PRESS

Santa Barbara / 1980

ACKNOWLEDGMENTS

"On His Way to Epley's Bike Shop Charlie Meets a Girl With 12 Dogs" originally appeared in *Playboy* magazine; copyright © 1969 by *Playboy*.

Other stories in this collection have appeared previously in the *10th Annual Year's Best Science Fiction, Cavalier, Oui, Writing From The Inside, The Santa Cruz Stain*, and the *Yes! Capra Chapbook Series*. Copyright © 1964, 1968, 1972, 1973, 1979 by James D. Houston.

The author gratefully acknowledges permission to reprint the material on page 23 from TAKE AN ALTERNATE ROUTE by Paul Pierce, Shelbourne Press, Los Angeles 90035, copyright © 1968 by Paul Pierce; the material on page 69 from MIAMI AND THE SEIGE OF CHICAGO by Norman Mailer, The New American Library, copyright © 1968 by Norman Mailer; and the material on page 105 from AS I LAY DYING by William Faulkner, copyright © 1946 by Random House.

LIBRARY OF CONGRESS CATALOGING IN PUBLICATION DATA

Houston, James D
  Gasoline.

  "A Noel Young book."
  I. Title.
PZ4.H844Gas   [PS3558.087]   813'.54   79-28425
ISBN 0-88496-144-3

CAPRA PRESS, P.O. Box 2068, Santa Barbara, California 93120

this book is dedicated
to Fred Engelberg
who taught me how to shift

GAS, coined by the Belgian chemist,
J. B. Van Helmont (1577-1644), was
derived from the Greek word *chaos*.

—*Webster's New World Dictionary*

Cars will always be fun.

—Robert D. Lund, Manager
Chevrolet Division of General Motors
in *The Los Angeles Times,* August 20,
1979

contents

The world's largest bulldozer is the 67 ton 67 feet high Le Tourneau Crash Pusher CP-1 driven on 6 tires 31 feet 5 in. in circumference. The most powerful ground clearer is the $200,000 125 ton Le Tourneau Electric Tree Crusher. It is 74 ft. 4 in. long, can clear four acres an hour.

THE GUINNESS BOOK OF RECORDS

An Occurrence at Norman's Burger Castle

It's good for business to have
a nice looking girl at the window.

— Norman

I

AFTER EYEING HER FOR A WEEK Charlie Bates knows one thing for
certain. Thelma is a vehicle addict. This, he figures, is why she
happens to be working at a drive-in, the same way a beer-addict
he once knew spent his nights tending bar, the way latent pyro-
maniacs join the fire department. Each night she drives away with
whoever has shown up in the gaudiest vehicle. And Charlie knows
that if he is to win her, he will have to come up with something
pretty wild.

At first that is all he wants, simply to win her. He hates himself
for even trying. He is at the point where he despises cars of every
size and color. They are making the world a miserable place to
live and making his own life an outrage. The parking, the ulcer-
ating traffic, the upkeep money. But this Thelma, the girl at Norman's Burger Castle, there is something about Thelma he will do
anything to possess. It is, at least in the beginning, an ambiguity
which, like the vehicles she adores, both attracts and repels him.

It began with her hands, one night when he rushed in looking
for a Baronburger to go. Preoccupied, Charlie was interested only
in quick service. He didn't notice the girl at the window until she
snapped the white bag open. Long fingers slid the wrapped burger
in. Something about that motion, something delicate, sensual, almost obscene, caused Charlie's face to burn. Looking up into
hazel eyes he saw a glance that could have been demure, or it could
have been an invitation to pleasure unspeakably bizarre. It de-
pended, he surmised, on how well you knew this girl. In gothic
script above her bosom he read the name, *Thelma*. She looked
around twenty. Red hair hung to her shoulders in a profusion of
shiny curls. Her skin dusted with light freckles, she had the sun-
fed complexion and guileless manner of a farm girl. Yet for Charlie her glance implied that this warm bag itself contained some

11

erotic secret, a note perhaps, naming the time and place of their rendezvous, or a dirty photo hidden between the buns. An unnatural craving overwhelmed him. His hand in his pocket, reaching for change, was suddenly embarassing. He grabbed the bag and hurried back to his car.

For half an hour he drove around, trying to decide what to do. He got back to Norman's just in time to see her pale leg step into a metallic blue '62 Chevvy sedan, truck tires lifting its rearend three feet off the ground. Its axle was painted metallic silver. The drop from back to front was so steep that as they drove past Charlie on their way to the street, Thelma and her escort seemed to be falling through the windshield.

Charlie winced with revulsion. The guy looked about twenty-two, exactly the kind of guy he always disliked in highschool, because there was always his hot car to contend with. Charlie had mistrusted cars even then. He could never really put his finger on why. He just hadn't been able to bring himself to rely on them for very much. Now highschool is a long way back, and Thelma surely has a metallic heart to match the chevvy's point job. Yet the sight of her driving out of his life made Charlie's stomach shrink.

The next night he sat outside in his Volvo a long time watching her take orders. When the crowd at the window thinned, Charlie made his move. His timing was bad. At just that moment Thelma took a break. An acne-chinned boy scribbled down his order. Feeling cheated, and already regretting his improbable pursuit of this girl, Charlie stood there waiting for his Knightburger and choc shake. Irritably he listened to the muzak:

> *La cucuracha, la cucuracha,*
> *Ya no quieres caminar,*
> *Porque no tienes, porque le falta ...*

"Jesus," Charlie thought. What kind of burger bar plays La Cucuracha?"

There was no time to dwell on the answer. Thelma had re-appeared. Back beyond the big stainless steel griddle she stood alone, staring through a plate glass window at the row of fenders and hoodpieces. Snug blue uniform, crown-shaped paper hat atop that Botticelli hair. He watched her eating frenchfries, one by one, in a manner that was both commonplace and decadent — with what could be simply the homespun gusto of a girl snacking in the kitchen, or an excrutiatingly suggestive ritual. Charlie observed her body in the same way. The fries probably explained that fullness of flesh across her belly and hips. Was it merely fat, or a

12

Renoir-like excess, promising luxurious excesses of passion? He had to find out. He hung around til closing time and watched her drive off in a waist-high Masseratti.

On subsequent nights he watched her drive off in a Model A with dunebuggy fenders and a Bentley grill, in a one-time schoolbus sprayed metallic copper and powered by two exposed V-8 engines, in a re-built '49 Hudson cut open to carry an enamelled plywood camper rig.

Charlie was so outclassed he could only chuckle in despair. To offer Thelma a ride in his unimproved '66 Volvo would be like offering a marshmallow to a starving man.

By a stroke of luck, on one of these afternoon he happened to be drinking a few beers and sharing his dilemma with a friend who worked for a big construction outfit, and the friend, trying to help, told him about all this equipment they were getting ready to scrap, the way they do every time they finish one job and start another. There was in particular one bulldozer the friend was familiar with. Charlie, listening to him describe this rig — the engine could use some work, a few tread links were loose, but all in all not a bad little dozer — Charlie was so filled with self-loathing and so crippled with desire, he could not prevent himself from begging his friend to help him get it.

The deal was arranged. Charlie picked up the bulldozer about three one morning, to drive it home while traffic was light.

At a railroad crossing a cop stopped him, wanted to know where he thought he was going.

"About five more blocks!" Charlie shouted down at him.

"What?"

Five fingers up. "Few blocks!"

"Can you shut off the engine, buddy? Can't hear a word!"

"I'd rather leave it running!"

The cop climbed up onto the metal tread. "I said shut off your engine!"

"It's a little hard to start this thing! You hear that funny snort in there? It's missing!"

"I don't care! Shut it off!"

Charlie let out the clutch, started to ease forward, causing the tread the cop stood on to roll.

"Hey! For Christ sake, buddy!"

The cop jumped clear, pulled out his pistol. He was jogging alongside the bulldozer yelling at Charlie to pull over.

Charlie ignored him, sped up to get across the track ahead of an approaching train. Red lights began to flash. The guard arms lowered. The cop halted, raised his pistol, aimed, thought better of it, ran back to his patrol car to call for help, then sat there while a hundred and thirty freight cars rolled through the intersection.

II

At Norman's a moat has been dug at the entrance. A simulated drawbridge crosses it. On either side stand the blinking orange castle gates, connected overhead by a neon sign saying NORMAN'S BURGER CASTLE in huge gothic letters. Beyond the gates, at the end of a long two-lane promenade, stands the castle itself, a one-story square building painted to look like silver-gray stones. Above the service windows rises a stone facade notched for archers. A flood-lit turret stands at each front corner, topped with permanently windblown plastic flags.

It is four nights later and nearly closing time when Charlie rumbles through the gates, over the drawbridge, and heads up the promenade to join the other glittering inventions that gather in Norman's parking lot.

An ordinary bulldozer would get no more than a glance from the crowd at Norman's. They would figure it for a work crew moving in to tear out something, install something else. But Charlie's is no ordinary dozer. The numbered stock cars making their final circuit of the castle's perimeter slow to a crawl as he enters. The lowered Mercs and molded Caddies revving up to fill the lot with fumes before departing for the night cut their engines back to an idle. Heads appear at the tiny windows.

Secluded in his garage, Charlie has spent the last four days decorating his new machine. Commercial art happens to be his line of work — freelancing to agencies, sign-painting, lettering, book jackets, things like that. Narrow brushwork is his specialty. In the case of the dozer, his sourness toward vehicles has spawned a perversely meticulous affection. Formerly rust-scarred and dusty yellow, it now shines with a paisley pattern of magenta, gold, jade and baby blue. Sperm-like figures wriggle around one another, snaking up the high exhaust pipe. Charlie has sprayed the tread links silver. The scoop is bright red, like an open mouth. A second tractor-type seat has been welded next to the driver's, and over these hangs a fringed green canopy, like a surrey's.

Charlie himself wears a buckskin coat, with long sleeve fringe, striped bellbottom pants, a blue workshirt and silver hard hat. On the way over here, trying to avoid traffic cops, while creeping

14

through night-time lanes of amazed and often snickering motorists, Charlie has been alternately pleased and disgruntled with himself. Ten times he was on the verge of turning back, or jumping out and abandoning his scheme entirely. Two things kept him going: 1. his goatish fantasies (*they pull into Charlie's garage. Driven to frenzy by the big machine, Thelma writhes like a belly dancer, tearing at his clothes, hungry for the skin of the man at the wheel.*) 2. a feeling, born in these last four maniac days, that a great deal more than Thelma is now at stake. Inside Norman's, moving slowly past a double rank of parked cars, like a tank commander reviewing troops, Charlie has a delicious premonition that tonight he will not only have the redheaded burger girl, but by having her he will somehow get his revenge.

He isn't quite sure against what. The smell, as much as anything else. The smell of Norman's Burger Castle is the perfect atmosphere for Charlie's mood. Approaching the turrets, he is surrounded by those salt-savory fumes he can never resist. He knows where they come from. From inside the shiny buns. That is all the buns contain. Fumes. Air. That is why the burgers give him gas, heart-burn, flatulence. He knows the meat is thinned, and tinted. And the shakes, if you let them sit for ten minutes, turn to sugary scum. But that smell, as if blown out toward the street by giant fans, with its promises of instant food, and instant fulfillment, the smell always brings him back for more. Oh, how it exploits his weakness, fosters this ruinous dependency. No matter what they do to him, the burgers are so goddam HANDY. He can't stay away from Norman's any more than he can rid himself of his car, or, for that matter, of this bulldozer. It is all the same. He needs the machine. And he loves the smell. And he detests the smell. And he drives through it recklessly now, a little bit, he imagines, like Napoleon, against improbable odds, starting north into Russia.

He pulls up near the window and, with exhaust pipe pucketing, sits there as if studying the overhead menu. He doesn't turn off the engine. He still has trouble starting it. He lets it belch and thunder, figuring he won't have long to wait. Thelma quits work in ten minutes or so.

Ignoring all the looks he's getting, Charlie hops down, saunters to the counter, and tries to catch her eye. An apricot redness tints her cheek. She is arguing with a man wearing a Japanese hapi coat and lounging pajamas.

"I ordered seven Baronburgers," the man says, "two Princeburgers and a Queenburger."

15

"You told me five Baronburgers."

"I meant seven."

"Why didn't you tell me seven?"

"The names confuse me. Why can't you call them large, medium and small, something like that?"

"Call them whatever you want. But when you tell me five, you get five."

"I want seven."

As she turns toward the Baronburger shelf, he leans across the counter, head into the window hole. "And who's responsible for this music? *Carry Me Back to Old Virginny,* for Christ sake! What kind of burger bar is that?"

When she doesn't answer, the man whines the lyrics at her:

Back where the cotton and the corn and taters grow...

Sliding his carton through the hole Thelma says, "You are responsible, amigo. That comes to five sixty."

Leaving, the man looks up at Charlie, a little guiltily, as if he has been caught at something. He has an enormous head of brown curly hair, hornrim glasses, a brown hitler moustache. He cocks his head, says caustically, "Some waitress."

Charlie shrugs, steps ahead, leans to speak through the opening. Thelma leans. Long amber curls fall past her ears, almost touching the counter. It is an oddly intimate moment, like popping to the surface of a pool inches from another dripping face.

He says, "Who was that?"

"That's Norman."

"Himself?"

"He's always coming around in disguises like that, to check on how we treat the customers."

Charlie turns to look at Norman again, sizing him up. Norman stands at the curb sizing up the bulldozer.

Turning back to Thelma Charlie says, "There's some truth in what he says, you know. The names? The music?"

She doesn't respond. She too is sizing up the dozer, staring over his shoulder in the direction of that plocketing rumble. She murmurs, "What would you like, Charlie?"

She throws him her ambiguous glance. Is it shy? Is it lascivious? A terrible certainty grips him. Thelma is finally just a very nice girl who happens to like to go riding in cars, and this grotesque contraption he has created is nothing more than a tribute to Charlie's own sick imagination.

Stalling, he scans the menu again.

"Oh . . . I guess . . . well, what the hell. Give me a Dukeburger."

"Better make it two Dukeburgers, Charlie. And some fries. Two or three orders. I just love those fries."

Staring past him, she says she'll meet him at the dozer in five minutes. He watches her walk to the wrapping table, watches her hips pull at the blue uniform. He is giddy, suddenly feeble. He needs strength, reassurance. He turns to see how his rig is doing. He sees Norman up in the driver's seat, fooling with the controls. The wide red scoop is slowly rising.

"Hey!" Charlie yells, "Get down offa there!"

Norman is pulling at the levers in front of him, at handles hanging over his shoulder. His hapi coat flutters in a light breeze.

Charlie leaps up next to him. "Hey, get away from here!"

"This your bulldozer, amigo?"

"Damn right, it is!"

"Well, get it the hell out of my parking lot! The noise is spoiling people's appetites. You're taking up too much space. I could put three cars in here, three family-filled cars, kids dying for Dukeburgers, double frosties!"

He yanks, he twists in the seat, revs the engine. "How the hell do you get it moving?"

"I'll move it," Charlie yells. "I'll move it! Just get outta the way for a second!"

"Look back where you gouged up the promenade with these goddam treads. My god, amigo, what are you up to anyhow?"

The scoop reaches its top limit and crashes to the ground, breaking off a chunk of curb. Norman's foot rams the accelerator. Frantically he tears at the levers, Charlie now struggling to shove him off the seat. Norman jabs something and the engine starts to cough. Two dark puffs snort up from the exhaust. Charlie lunges for the choke. Too late. A sputtering shiver, and the engine dies.

A strange silence settles over the burger castle. The muzak is caught between songs. In the stream of passing traffic there's an instant's lull. Nearby engines idle quietly. Late burgers hang half-eaten between lap and chin.

Through loudspeakers a raspy voice says, "Number 48, your order is ready. Number 48, please." The music resumes.

Charlie says, "You asshole."

"What?"

"Sometimes it takes me thirty minutes to start this thing."

"I hope it doesn't take that long tonight."

Thelma appears next to Charlie's knee, smiling up at him. She has exchanged her uniform for floral patterned harem pants and a green velvet vest, nothing underneath, its sides joined by a small gold chain across her navel. As she hands him up the warm bag full of fries and Dukeburgers, Charlie can't help peeking through the arm holes, richly embroidered, cut low and loose. An agonizing peek. The face she wears is so sunny, so rural, even child-like, it seems to say that only a deviate could take this outfit the wrong way.

Bitterly Norman says, "Is this your boyfriend, Thelma?"

"We were going for a drive."

"A drive, is it?"

Norman's mouth twitches. His eyes scan Charlie enviously, yet somehow like a father's looking over his daughter's date. Charlie would like to speak to him privately, feels they have a lot in common.

"Well, you have five minutes to get this show on the road."

"Five minutes," Charlie blurts. "Jesus, Norman. I bought a couple of burgers, didn't I? Sack of fries? Don't I have a right . . .?"

"Five minutes, amigo. After that, I call the gendarmerie."

"You're closing up, aren't you? What difference does it make?"

Norman cinches his hapi coat and jumps lightly to the asphalt, picks up his carton of burgers. They watch him stride out of sight around one of his stonework corners, zoris slapping, pajamas flapping losely at his ankles.

"Don't mind him," Thelma says. "Show me around this *machine.*"

Hopping down, Charlie joins her by the scoop. She examines its blade. She bounces a knuckle along the metal links. She smooths a hand across the jades and baby blues of Charlie's paisley paint job, the same way she slips burgers into bags, except now there is no misreading her gestures.

"How do you get *into* this thing?"

"Here. I'll help you up."

With hands around her bare waist, Charlie lifts her to a sitting position on the silver tread, finds his thumbs just below her bosom, and lets them linger there, testing the curve. She likes that. Her curls graze his forehead. Her fingers slide along his neck. Relieved, aroused, exuberant, Charlie springs up next to her, and with one hand on the elbow, one under the vest, across the velvety back, he guides her beneath his canopy.

Seated, he reaches down to switch on his tape deck. Blues organ and heavy brass section pour from overhead speakers. Into their ears the organist shouts

18

You never told me, baby,
That your legs were big and fat.
You misinformed me, baby,
And you know I won't stand for that.

Rocking with it, Thelma glances approvingly at Charlie's side-burns, his buckskin lapels. Her freckled cheeks flush, her eyes shine with a virginal admiration, as if his leather is the first she's touched, his vehicle the only one she's seen. She runs an amorous hand up under his hardhat, into his hair. "Charlie, this is the sex-iest thing in the *world*," leaning then, in the glare from all parts of the burger castle, to kiss him warmly on the lips.

Charlie has this part of it all planned out. With one arm he pulls her close, with the other he reaches up to a canopy corner, pulls a cord. The back curtain flops down. He pulls another, and one side curtain drops, cutting off the direct light from the kitchen, and the view from most of the cars remaining in the lot. Thelma is reach-ing inside his coat. Her lips against his chin are soft, eager.

"Let's *go* someplace," she says.

He reaches for her silky thigh.

She murmurs, "Drive me *around*, Charlie."

He had this part planned out too. The getaway. Now he needs some time to think. And there isn't much. Thelma's fingering the buttons on his shirt. The trouble with dozers, they're too heavy to start with a push. He could try the jump wires in his tool chest. Or he could call a towtruck. Or perhaps he could beat Norman's head against the engine. Yes, that might be the answer. Awaken all those drowsy connections with the back of Norman's head.

Expecting the worst, but trying for the nonchalance of an old pro checking out his equipment, Charlie lets his eyelids droop knowingly, turns the tape deck down so he can hear, and presses the starter.

It grinds. The engine bucks. Some fries fall to the floor. It grinds and it grinds and it grumbles and it groans, and Thelma is sucking on Charlie's ear. Her hair falls against his neck, each strand and ringlet a tiny tongue enflaming him.

He lets go the starter, clutches her leg again, maddened by the slide of flimsy cloth across her skin.

Taking her lower lip between his teeth, nibbling, he says, "Lis-ten, Thel. Just in case I *don't* get it started . . . I mean, I know I will. But just in case I don't, how would you feel about *walking* over to my place. It's only a few blocks from here. I've got some beer in the icebox and this bag of Dukeburgers. We'd be all set."

19

He feels her stiffen. "Oh Charlie, I wouldn't want to go any-place without the *dozer*."

A loud click. Over the loudspeaker a gravelled voice shouts, HAVING SOME TROUBLE OUT THERE, AMIGO?

Charlie hits the starter again. It groans. The engine blaps, as if about to catch, but doesn't. It grates and it grunts and it groans and it grinds, and Charlie is cursing everything in sight, including Thelma. The weakness in her. Falling for a rig like this. He would like to violate her somehow. She deserves a little rough treatment.

"Thelma," he declares, "I am calling a cab . . ."

TIME TO MOVE IT OUT, BIG FELLA!

Charlie throws aside one curtain. "Oh shit, Norman! How can I move it when I can't even get it started!"

Charlie can see him in there next to the mike, giggling and look-ing around at the cleanup crew to see who is digging his routine. His fright-wig is gone. His natural hair is black, thinning, slicked straight back. Charlie figures him for about forty. He still wears the hornrims, the brown hitler moustache. His voice booms all over the lot.

THAT'S NOT MY PROBLEM, FOLKS. TIME'S UP. I AM CALLING THE GENDARMERIE, THE CARBINIERI, THE FUZZ, THE HEAT, THE BLACK MARIAH. THIS IS CAPTAIN MIDNIGHT SIGNING OFF. Click.

He whirls triumphantly, heading for the phone. Charlie jumps down, sprints to the window.

"Hey!"

A sign drops in front of his face: SORRY. WE'RE CLOSED.

"Goddam it, Norman! Can I talk to you for a minute?"

On the other side of the glass, Norman is silently mouthing words that seem to say, *I can't hear you. Please speak a little louder.*

Charlie shouts, "You stupid sonofabitch! You're the one who killed my engine in the first place!"

Norman reaches below the counter. A blast of deafening music rattles the speakers:

> *Swing low, sweet chariot,*
> *Comin for to carry me home.*

Behind the glass Norman mimes a softshoe.

Enraged, Charlie runs back to his dozer, leaps aboard. He hits the starter so violently that for some reason it nearly catches. Thelma hugs his arm in anticipation. He hits it again, and the great engine bellows to life, vibrating like a jackhammer. Thelma

20

squirms in her seat. "Oh *Charlie!*" She kisses him, pushing her tongue between his determined lips. In his absence she has been munching fries, and this taste, this smell is some final trigger for his passion. He will have Thelma. But first, by God, he will have Norman and his whole intolerable castle.

"Hold on!" he yells. "I'm going to get that sonofabitch if it's the last thing I do!"

He raises the scoop til it's right in front of their eyes.

Through the mike Norman shouts, YOU WON'T DO IT, CHARLIE. YOU'RE CHICKENSHIT!

"The hell I am," Charlie yells back, and the dozer leaps forward, smashing through one corner turret.

He backs off. Thelma says, "It's so *powerful*, Charlie." Just before she drops the other side curtain, he sees the cleanup crew, aprons flying, running for their lives. Thelma unsnaps the tiny navel chain. The sides of her vest fall back.

A brutal kiss from Charlie, a brief fondle that makes her cry out, and he is heading for the other turret. Norman screams, YOU DUMB BASTARD. YOUR AIM IS ALL OFF. YOU WON'T EVEN HIT THE FUCKING THING!

But Charlie hits it head on, barrels through. A two-by-six crumples his exhaust pipe. Plaster and lath come crackling down, and he is turning around, backing up now for the final pass.

Thelma is hysterical with ecstasy. She has ripped off her harem pants and is trying to grab the controls, forcing Charlie to wrestle for them. She ends up in front of him, marble-bottomed and bent slightly forward to handle the levers. He has no choice but to take her from behind. He holds her by the shoulders — his steering wheel, his shield — as the big dozer lurches toward the castle wall.

Hand on the throttle Thelma's yelling, "Faster! Faster!"

Through the loudspeaker Norman screams, C H A R G E. C H A R G E.

Charlie roars, "Now, Thelma! Now!"

Just as the speakers shatter with their overload, she ploughs through the wall of painted stone, on into the women's john, through that wall, bursting out among the burger shelves and relish jars, toppling pillars of plastic cups, exploding great jugs of mayonnaise and ketchup, halted finally by a heavy eight-foot freezer which the dozer can't move very far until the scoop is lowered.

The dozer bucks against it. Thelma, suddenly powerless, has let go the levers. Impaled, she falls back against Charlie, pinning him in the driver's seat. He struggles to lift her, to reach around

21

and get the rig moving. But he too is weakened. They are an eight-legged spider, arms and legs flailing, when Norman crawls out from under his counter, moustache askew, but otherwise untouched by this onslaught.

Brushing dust from his pajamas he scrambles up onto the tread, throws the curtain back. Charlie, vulnerable, unable to move, and seeing the crazed look on his face, is terrified. "Hey, Norman . . . Look, man . . . I . . ."

But Norman is shoving in front of Thelma, shouting, "Christ Almighty, amigo! You paralyzed or something? Let's get this show on the road!"

Dropping the scoop, he gets a bite on his freezer, flips it. This clears the way. He backs up and crashes out through the far wall. A shred of lath grabs the canopy, tearing it off, taking speakers and curtains. Norman doesn't stop. He is singing *Swing low sweet chariot* and clattering down his promenade. He rams the first gate, on the outside of a U-turn. Then, circling back into the empty lot, he pauses.

They are sprinkled with flickering orange light. Norman shakes Charlie. "C'mon, get with it! You too, Thelma! Get off of Charlie and put your pants on!"

Charlie says, "Norman, I . . ."

"If you're trying to apologize, don't."

"I was just going to say I'll flip you for the second gate."

"No contest, amigo. It's all yours. You deserve it."

Generously Norman steps out of the way. Charlie moves Thelma onto the other seat and slides forward, taking control again.

"Anything you want to say before we go, Norman?"

"Yes. Yes, there is." He jumps onto the rear platform and screams, "I HATE BARONBURGERS! I HOPE EVERY REMAINING BARONBURGER ROTS IN THE DEEPEST PIT OF HELL!"

For a moment he stands there, listening to this echo back over the wreckage.

Charlie says softly, "I know how you feel."

Norman squats behind the seats, purged, squinting warily at their last target. "Let's get out of here."

The scoop rises with a painful screech, bent by its impact with the freezer. The exhaust pipe is folded double now, the paint job has been mutilated. Ragged green banners hang from the two remaining canopy supports. Charlie uses this new scoop angle to land the second gate a high glancing blow. It shorts out the orange blinkers and the row of elevated lights. As he rumbles over the drawbridge, with the lot in darkness behind them, the gothic sign crashes into the moat. As they swing into the outside lane he hears Thelma next to him saying, "Get your hands off me. Charlie, can you make Norman quit . . ." Then her voice is muffled. He doesn't turn to see why. At ease with the world, his hands resting easy on the levers, and his hardhat cocked, Charlie sits tall in the tractor seat, hums their victory anthem, and thunders on into the night.

✥⳥●⳥✥

. . . large areas of the original coastal strip were covered over altogether with macadam, white condominium, white luxury hotel and white stucco flea-bag. Over hundreds, then thousands of acres, white sidewalks, streets and white buildings covered the earth where the jungle had been. Is it so dissimilar from covering your poor pubic hair with adhesive tape for fifty years?

NORMAN MAILER, *Miami and The Siege of Chicago*

✥⳥●⳥✥

On His Way to Epley's Bike Shop Charlie Meets a Girl with Twelve Dogs

TEN O'CLOCK ON A SATURDAY morning and here comes Charlie in his big VW bus, red white and blue like a mailtruck. He's crossing town to a bike shop to have his gear cable fixed, exhilarated by the brand-new feel of this day, thinking how rain has rinsed it clean, and vaguely watching the road for hitchikers. He likes to pick up hitchikers on such a day. He'd like to pick up a highschool girl. He sees so many on the roads now, wrinkle boots, ranch coats, straight long hair. Something about the long hair gets him, some flash of recklessness. He'd like to flirt a little, seduce one maybe, if he could figure out a way. He stretches to see his face in the rearview, pushes back his forelock. His wife is out of town for the weekend. His bus is equipped for random outings, mattress in back, tin skillet, kerosene stove, and at the moment one bicycle with cable wires dangling. As he rounds a corner ten blocks from home he sees the girl he might be looking for sitting on a curb.

She hovers over a box full of puppies, arranging them and reaching out to stroke the hair of a white, black-spotted Dalmatian mother. As Charlie passes, the girl extends one pale and leisurely thumb, without looking up. This in itself attracts him. He stops. The Dalmatian climbs in back onto the mattress, among spokes and handlebars, the box of puppies is squirming on the seat between Charlie and the girl.

She's around seventeen, maybe sixteen, wears calf-high moccasins and a skirt wide as a hand towel, so all but a few inches of her smooth legs are visible next to his gear shift. Plum-colored shawl, square-rim specs, red hair to her waist but drawn back by a plum-colored ribbon, grandma style, and her face scrubbed clean. No make up. Her smile has a chaste, benevolent wholesomeness Charlie associates with nuns and candy-box illustrations and certain passionate women who are holding it all inside.

"You going to the edge of town?"

"I have to stop at this bike shop first. But if you don't mind waiting in the car a minute I can take you right to the highway. Where you headed?"

27

Her voice is soft and breathy, ethereal, hard to follow because she speaks so slowly, as if trying to recall how each sentence is supposed to go. "Up to Mendocino actually . . . to visit a friend of mine . . . who might have a baby at any moment."

Charlie glances at the wriggling pups. "Mendocino is five hundred miles away."

"Yes . . . it will be very . . . difficult for her. She already has three children . . . This will make five . . . if it's twins."

"Well," Charlie says, "it's a good day for travelling."

With eyes clear and unblinking, with the same benign and guileless smile, the girl regards him.

"Yes, I'm glad . . . I want the dogs to enjoy it."

Charlie has to listen hard to hear her. In the back the Dalmatian whines, straining to jump the seat and join her puppies. "She's highstrung," the girl explains. The pups themselves are yipping and squealing and climbing over one another, heading nowhere, eleven snouts, twenty-two ears, forty-four black and white and gray and spotted legs.

At Epley's the girl covers most of them with one arm and reaches back to hold mother by the neck while Charlie pulls his bike out. He's wearing his long-sleeve paisley shirt — yellow cuffs and swirling turquoise, his wide-grain hip-hugger corduroys with broad black belt. He figures he must look pretty good wheeling his bike up the concrete path.

Epley waits in the doorway. His black hair is the inverted base of a triangle head. The inverted apex, his chin, is nearly black too, with stubble almost too tough to shave. Charlie feels sorry for the skin below his jaw, all the scrapings that have made that skin like a scraped bicycle patch. It hangs a little loose. He's nearly fifty. Black frisco jeans. Black t-shirt.

Thick glasses distort Epley's huge black eyes. He looks startled as he tells Charlie the trouble with his gear shift is inside the wheel, not in the cable, and he doesn't have the part in stock to fix this model. Charlie follows him inside to the catalogue.

Epley's shop is the front room of his house. In one corner lightning streaks crackle over seated bluish figures, sparks of static through the loud moderator's pixie smirk:

> "Sixty seconds left, folks. Big Clock counting it
> down. Your mother was here an hour ago. Is it real?
> Is it rumor?"

Worn sofa, six reading lamps, old bicycles, one lawnmower, three electric heaters he's working on, scattered spare parts, metal desk piled with tags and bills, against one wall an old glass-sided adding machine, and next to the sofa a tape recorder into which Epley sings from time to time, since he once tried to break into show business. Charlie learned this on his last trip to the shop, three years ago, when Epley played a few yards of himself singing "Joshua Fit the Battle of Jericho." Charlie can't tell whether Epley remembers him or not.

He watches gears and rods clank inside the adding machine as Epley computes ahead of time how much the repairs will cost. It's something under four dollars, and Charlie is moving toward the door, anxious now to drive the redhaired girl up the coast a few miles himself. At least to the next town. He has no clear idea of what he will say or do. But maybe something will happen. He has plenty of time.

Epley says, "Do you realize what they're trying to pull?"

"Who?"

"I can't even put bicycles on my lawn any more."

"How did that happen?"

"Let me show you something."

From between the cushions of his old sofa Epley takes a sheaf of legal-size paper, flipping pages too fast for Charlie to see the title.

"Just look at this, this sentence right here . . . *pertains to all businesses regardless of location or gross profit.* I call that dictatorship, buddy. That's what it's coming to. Thirty-seven years in this town, just to watch it turn into a goddam dictatorship."

The magnified eyes widen, as if witnessing some atrocity. He sways, shifting weight from foot to foot, whirls suddenly to face the set.

> "Big Clock rewinding, folks. Contestants changing places for Declaration Three. The man you're about to hear is the brother of a former ambassador to one of our NATO allies. Listen closely now. Is it real? Or is it rumor?"

Charlie makes his exit. He plans to use his bike gearless til the part comes in. Seeing that he might appear wishy-washy wheeling it back to the car, he starts to shrug his shoulders and make some comic face. But the girl isn't watching. She's hunched over the pups. Epley's angry voice is right behind him.

"Hey listen! Do you realize what would happen if that building over there burned down?"

Charlie doesn't answer.

"They could put a parking lot in there, which means this other fellow, two lots away, might have to paint his side window. Just cover it up!"

The big Dalmatian is barking, still trying to climb the front seat. The girl decides to leave her puppies and sit in back. She steps out, ignoring the two men.

Epley has started to parade along the sidewalk. "Over five hundred people signed the petition," he shouts. "Me and Mrs. Spragg did almost all the work singlehanded. Living dynamo, that woman, nearly seventy. Here, lemme hold that bike for ya."

While Charlie opens the side doors and they shove the bike in, the Dalmatian finally leaps over the seat. In her excitement she tips the box, and all eleven puppies come tumbling through the still open front door, out into the sidewalk.

The girl slowly kneels. Three are waddling up the walkway. Charlie moves to grab them. At that moment the Dalmatian unaccountably bounds past him, across the grass and through Epley's front door.

"Hey. That black and white dog of yours just ran into the shop!"

"It isn't my dog," Charlie says.

Epley clutches black thighs, stares wildly through his glasses, like a man surrounded by savages, first at Charlie, then at the girl where she kneels ministering to her pups.

"Waddya mean, it isn't your dog?"

She looks up at him, her lips spread slowly into that nun's smile. Epley grimaces trying to hear. "It's . . . my dog. She's . . . very highstrung."

Charlie catches the wandering puppies. "Here, hang onto these. I'll try to find the dog."

"Oh . . . thank you . . . thank you . . . very much."

There's so much space between her words, he has crossed the lawn before she finishes. Epley's already inside. Charlie hears him yelling in back. Lightning streaks punctuate the voice, disfiguring a blue-white row of faces:

> "You're wrong, Mrs. Notley. There's the buzzer,
> and Big Clock says everybody's wrong this time.
> It wasn't a rumor at all. Canadian Riley is still
> alive and she'll be onstage here in just a couple of
> minutes."

Charlie kicks old milk cartons out of the way to get through Epley's kitchen. On the table lies an overturned kingsize Cornflakes box with top torn back, some murky pyrex bowls. Pans are stacked on the stove and drainboard. Cupboard doors stand open.

30

Epley's in the bathroom wrestling with the dog, who has rooted out some stained and grimy underwear.

"Look here, mister! Tell this goddam dog of yours to let go of my shorts!"

Charlie tries to grab her, but she's snarling now. He draws back, afraid she'll claw his corduroys.

"Why don't you let her have the shorts?"

"Are you nuts?" Epley yells. "Are you crazy, pal? What the hell is going on around here anyway?"

"I'll have to get the girl. It's her dog."

Back through the cluttered kitchen, Charlie's heading for the front door when he finds her sitting on the floor, back against the adding machine, holding the box of puppies in her lap, watching the screen. Her shawl has fallen open and underneath it Charlie sees nothing but white skin from neck to belt, and below the belt about eight inches of denim skirt before her legs start.

She isn't self-conscious about this. She doesn't even look at him.

"You'd better call your dog. She's chewing up Epley's shorts."

She stands, draws the edges of her shawl together, walks into the kitchen, lifting moccasined feet as if picking her way across a floor covered with guitars. Her steps make no sound. When she's gone, Charlie squints in amazement at the row of puzzled faces in the corner.

In the back the dog snarls. Epley's yelling, "Goddam you, black and white dog bastard. Let go! Will ya let go?" Charlie hears him yelling at the girl. "Hey, what the hell *is* all this?"

Then the dog shuts up, and the girl's in the kitchen, dragging her by the collar. Right behind comes Epley, stuffing torn shorts into his back pocket, and trying to step around them, shifting from foot to foot as the girl pauses in his doorway.

"Jesus Christ, lady. Why'd you have to bring all these dogs into the shop?"

She turns to him. "Do you happen to have any oats?"

"Oats?"

"Dried oats." The smile again, everybody's grandmother, everybody's saint. One hand holds the dog, who strains to reach Epley's back pocket. The other holds the edges of her shawl. Epley's enormous eyes stare at the white hand against the plum colored shawl. "It's what I feed the puppies after Nell nurses them awhile. Just a big cupful, and anyone who didn't get enough milk can fill up on oats."

Epley starts to massage his scalp, digging grimy fingers into the black inverted triangle of hair. He strides to his desk, kicking old sprockets with his heavy boots. One broad sweep of his arm clears half the desk of papers and tags. They flutter to the floor.

"Jesus Christ Almighty!"

He grabs the handle of his lawnmower, shoves it across the room and right out the door, so that it clangs down three steps and over-ends into the grass.

"Look at that lawnmower, for Christ sake. Now what am I gonna do? Lady wants it fixed by three this afternoon. And look at it now!"

All this noise scares the pups. They're yipping and wiggling under one another to escape it. The girl joins Charlie at the adding machine, squats, reaches in with one hand to knead and reassure. The way she squats, with knees spread, it's clear to Charlie she's wearing nothing under that short skirt either.

Nell has started barking again. On the screen small faces open wide in loud guffaws, cameras cut from mouth to mouth. Epley has switched on his tape recorder, his own warped baritone fills the room: *Joshua fit de battle of Jericho, Jericho, Jericho.*
  Joshua fit de battle of Jericho,
  And de walls come tumblin down.

And now he's rushing around grabbing bikes that lean against his walls. He runs them out the door one by one and they bounce down the steps, piling up around the lawnmower.

"Look at that!" he shouts. "Look out there on my lawn. A week's work, all shot to hell! What am I supposed to do now?"

Nell strains and snaps at the torn back pocket of his frisco jeans. The girl murmurs, "Nell. Nell, sit down."

"Maybe we'd better get going," Charlie says. "Why don't I carry the pups out to the car. Nell might follow them."

Epley yells, "Hold it! Hold it right there, buster! I got a week's work stacked up on that lawn, ya know — busted wheels — and a mower that'll probably never go again."

"I guess I could help you bring all that stuff back inside."

"What good would that do?" The grizzled face contorts.

"Didn't you say you couldn't park your bikes on the grass any more?"

"Fuck em!" Epley shouts. "Fuck the bikes! Fuck the rules! Fuck em all!"

Nell tears loose from the girl's grip and lunges at Epley's rear, rips the back pocket off his jeans, mouthful of old shorts and black pocket patch, bucking her head with the prize.

"Here. Gimme that, you goddam dog bastard. Gimme that!"

Epley dives for the dog, hunched and hairy-armed, chasing,

lurching. Charlie grabs the puppies, hurries toward his red white and blue VW.

"Hey you, where dya think you're going with that box? What about my lawnmower? What about all these goddam bikes? Goddam it, dog, c'mere with my underwear. Make this dog bastard slow down a minute, lady. What the hell *is* this anyhow? Look at my lawn!"

Following Charlie Nell bolts outside, Epley right behind. Charlie opens the front door, sets the pups on the seat, opens the double side doors in time for Nell to bound up onto the mattress.

"C'mon," he shouts to the girl, "let's get going."

She appears in Epley's doorway like a sleepwalker, surveying the scene — the bike-littered lawn, the dog-filled bus, Epley sprinting the concrete walkway — with her benign smile, as if to raise arms and bless it all.

From behind, around her shawl and out the door, come Epley's phlegm-splattered lyrics, into a second chorus, the voice louder now, and trembling, as if with anger, as if Epley is a fevered evangelist, aching for new signs of faith such as this:

> Oh de lamb-ram-sheep horns began to blow,
> And de trumpets began to sound.
> Joshua commanded de chillun to shout,
> And de walls come tumblin down.

"What about my bikes? What about my lawnmower?"

Charlie starts the engine. The girl drifts toward him, not touching the ground. At the curb Epley meets her, astonished and magnified eyes gaze at her square sun-flashing spectacles. Fingers wriggle at the ends of his arms.

"What about my underwear?"

She slides in. As Charlie pulls out into traffic, Epley sprints to his tangle of bikes, frees one many-geared speed racer, hops on and starts across the grass, bouncing the curb. With knees pumping furiously he catches Charlie at the first light, pulls alongside, yelling, "Hey, where do you think you're going, for Christ sake?"

Red light snaps green. Charlie guns it. But in this traffic top speed for his VW bus is 30. Epley keeps up, pedalling along next to the window.

"You sonofabitch with your bastard dog! Gimme back my underwear!"

Charlie shuts his window, switches the radio on, up full:

"The next caller is Miss Rosalie Dimond over there

in Pacifica. Our operator is ringing her now. I can hear it buzzing. (*CLICK CLICK*) Hello, Rosalie? Rosalie? (*CLICK*)

At the edge of town, where the street becomes a highway, traffic's still just heavy enough to hold Charlie back. He can't pass or pick up speed. Epley's right behind him. Charlie's afraid to drop the girl here, worried about Epley's harassment. Yet there's no point in driving her up the coast, not while Epley's tagging along. Nell is howling now. Some puppies have spilled out onto the floor. Charlie has to watch his feet on the pedals. The girl leans to gather the pups. Her shawl falls open again, white breasts hanging, and she lifts her eyes to look at him with her saintly smile, while outside the window Epley's stubbled, sweating and tortured triangle of a face yells over the radio and Charlie's engine, "Turn around. Will ya turn around? You think I can leave my shop open all day with nobody to watch it, and bikes scattered all over hell? I've gotta get back to town, for Christ's sake! Turn this rig around, buddy. Goddam it, my legs are getting tired. What'd ya pick a track like this for *any*way?'"

Town is two miles behind when Charlie spots a narrow road off to the right. It's unmarked and unpaved, but graded, and it winds steeply into mountains that lean back from the sea. He turns suddenly, and Epley goes shooting past. In second gear Charlie starts to climb, grinning. He turns off the radio. In his rearview he sees Epley circle back, struggle to pump the grade. Finally, rounding a high banked curve, Charlie sees him far below, stopped, straddling his bike and flapping both arms like a man warning traffic at an accident. He picks up what appears to be a rock and hurls it in Charlie's direction.

Then Epley is out of sight. Charlie keeps climbing. After a while the girl says, "You said you were going to drop me at the highway."

"Well I don't mind driving up the coast a ways."

"Is this the road to Mendocino?"

"No. Just trying to shake Epley."

"Will that take very long?"

"He's probably already started back to town. We'll just follow this road a few minutes more, then head down to the highway again. Don't worry."

"Oh, I'm not worried."

Her voice in fact is full of trust, total repose. A couple of curves later she says, "But could we stop somewhere? For Nell? She's very sensitive. She gets carsick . . . on curves. I'd rather just stop somewhere and wait . . . if that's all right. Would you mind stopping right over there?"

She points to a gravelly wide space next to a knoll covered with long grass, a high mound shining green after the long rains. Charlie parks. She climbs out, opens the side doors and reaches in to fondle Nell, rub her sides, lets the dog lick her face and neck.

Nell jumps out to scamper around her legs, the girl takes the puppy box and meanders toward the knoll. She moves in that slow, soundless way, as if each step is a thing in itself, to be savored — the only real step anyone has ever taken. This appeals to Charlie. When she wades into the grass, starting to climb, he follows, imitating her pace.

The grass stands two and three feet, topped with pale tassles hung from capillary branches, all translucent now. Charlie drags his hands through the tassles, and his heart swells with amused and generous gratitude, a surge of warmth for the man who forced them toward this hillside retreat.

At the knoll's top he finds her flattening a grassy oval. She smiles again, the smile that purifies every act, somehow makes it holy. She removes her shawl, spreads it out behind her, and she's lying there in a plum colored nest, miniskirt and moccasins, with arms stretched back and the sun shining two silver squares off her specs.

"I just love the sunshine. So does Nell."

Charlie looks around. No other cars on the road. No houses. Not even a cow. Just this spring mountain sloping to the sea, wooded in spots, broken by gullies and other little mounds and knolls. He remarks how he loves the sunshine too and starts to remove his turquoise paisley shirt.

"What's your name?"

"Charlie."

"Mine's Maude."

"I've never known anybody named Maude."

"Hey Charlie?"

"Yeah?"

"Will you sort of dump those puppies out all over my chest and stomach."

"What?"

"I love to feel em crawl like that. They're so furry."

"Don't they scratch?"

"They just like to squirm and cuddle."

Charlie dumps the pups onto her stomach and chest, and the big

35

Dalmatian is circling like a cowhand. Charlie starts to take off his boots.

"I could spend all day up here," Maude says. "Just lying around in the sun."

"So could I."

"You want a few puppies? Let em crawl around on your stomach too?"

"Sure. I guess so."

Dreamily she rolls her head toward him and half rolls her torso in a gesture of offering. "Here, help yourself."

Charlie is leaning toward her flat white belly, reaching for one warm handful of grey fur, when he hears behind him the scratch of loose rock, he hears heavy breathing. He turns to see Epley's black hair clearing the knoll's far side, then the agony of Epley's face, running with sweat. His great eyes blink insanely behind dripping lenses.

It's a steep short dropoff over there and Epley is clutching for handholds. "Hey! Jesus Christ Almighty! What the hell kind of a deal *is* this? Give a guy a fair shake once in a while!"

Charlie jumps up. "You get the hell out of here, Epley! Go on! Get back to your bike shop!"

Nell has spotted him and comes tearing through the grass like a maddened guard dog, neck stiff, and growling.

Maude pays them no attention. Between her eyes and the sun she holds one puppy, tickling it, cooing and snicking with puckered lips.

"Can you do that?" Epley yells. "Can you give a guy a fair shake? I was doing okay til you turned up this godforsaken turnpike. I blew a brand new tire before I'd gone a hundred yards. What am I supposed to do now? Hey, call off this dog, will ya? Get away. G'wan. Dog bastard. Will ya leave me alone?"

"Sic em, Nell. Sic em," Charlie says. "Get Epley."

Nell has forced him to the precipice. Epley loses his footing and steps backward, sliding down the short rocky wall he just climbed. Nell pursues him, and Epley in his loggers boots, his burr-prickley frisco jeans and back pocket gone, takes off galloping and yelling down the hillside.

"Hey! Tell this dog bastard to cut it out. Can you do that, for Christ sake? Get away from me, black and white dirty dog sonofabitch in the grass! Hey! Hey! What do you people think this is — a picnic area? I got a flat tire and my front door wide open. What am I supposed to do *now?*"

"Hitchike," Charlie yells after him. "Ride on the rims. Do whatever you want, but get the hell away from here. Sic him, Nell. Get him good."

He watches til they disappear in a grove of trees. Then he hears her soft voice. "Charlie? Charlie?"

He turns to find that she's removed her skirt. Just high moccasins now, the spectacles, and eleven puppies swarming over her body.

"Charlie, doesn't the sunshine make you sort of . . . you know . . . feel like doing something?"

"What about Epley and Nell?"

"I mean before they get back."

"I'll put the puppies in the box."

"Don't you like my puppies?"

"I love your puppies."

"You're not oldfashioned, are you? I mean, you're not hung up on some cornball style of . . ."

"Not me, Maude, never let it be said."

Charlie has stepped out of his corduroys and is kneeling next to her when he hears the engine of his VW starting down below.

"Did I leave my goddam keys in the car?"

"C'mon Charlie," Maude murmurs, "hurry up."

He runs to the knoll edge, sees Epley behind the wheel and Nell leaping at the window wing, hurling herself against the door.

"Hurry up, Charlie. This sunshine is turning me on."

He sees his bus swing out of the gravelly wide space. In his shorts Charlie's sprinting down the hill toward the graded road. But Epley swings it again, cutting two tracks through the deep grass, and he's heading up the slope, straight for Charlie, with Nell alongside, leaping and barking with hate.

"Epley, you stupid sonofabitch!"

Behind the windshield Charlie sees the wild eyes, more startled than ever, terrified. He jumps aside, and as the car careens past him he grabs the open window, one foot on the tiny step, head next to Epley's head.

"You're going to wreck my car!"

"What do you want me to do?"

"Stop! Stop it right here!"

Just as the front wheels clear the rise, Maude is coming slowly to her feet, red hair pouring over white breasts and shoulders, puppies drip and fall to the grass like some fur coat falling apart around her. Epley brakes in panic, his head hits the window, Charlie tumbles into the grass.

Maude catches two falling pups, lifts them to her neck like a muffler and begins walking toward the car with a pleased smile, some priestess receiving an expected pilgrim. Epley's foot slips and the car, rear wheels still on the incline, starts to roll backward.

"Brakes!" Charlie yells, "Hit the brakes!"

But Epley is paralyzed. The bus picks up speed, rolling silently. Nell stands watching now. Maude whispers Charlie's name but he doesn't turn. He sits in the grass and watches his red white and blue bus roll across the road and bounce off the far side, into more grass, a steeper slope that drops about fifty yards then gradually levels out before it slopes up the other way. Through the bottom of this draw runs a rocky ditch, the bed for a water trickle slipping seaward. Charlie watches his rear wheels drop into this ditch, he hears the crunch.

There's a long silence — Charlie clasping his knees in disbelief, Maude above him with two neck-warming pups, other invisible puppies strangely quiet in surrounding grass, the only sound a thin piping whine from somewhere deep in Nell's throat as she too waits for the distant bus to do something.

At last the horn sounds, a long waaaaaaaaaaah fills the little valley. Another waaaaaaaaaah, and waah waah waah, and Epley's head out the side window. "Hey!"

Waah waah waah waah.

"Hey, can you hear me?" Waaaaaaaaaaah. "Can you gimme a hand down here? Some kind of shove or something?"

No answer.

"Hey!"

Waaaaaaaaah. Waah waah waah waaaaaaaaaaaaaaaaah.

"Hey, you people gonna stand around til lunchtime? I'm stuck down here in the bottom of this gulley, for Christ sake!"

Nell barks back but without much conviction; he's too far away to interest her. She drops her head, rooting through the grass to round up her litter. Maude drops to hands and knees, lazily crawls around imitating the dog, shoving puppies with her nose toward some center point. Like Nell she wags her tail each time she locates a pup, and Charlie watches her round buttocks wriggle in the sun. He studies her nippled, ground-pointing cones, and all about her the backlit thicket of slender stalks, translucent, seed-heavy tassles bending to tickle milky skin.

He can't stand it any longer. He rolls over on all fours and scuttles toward her, planning to take her from behind, by surprise.

Waaah waah waah comes Epley's manic horn from the gulley. Charlie ignores it.

38

"Hey, what's going on up there? Jesus Christ, can't you give a guy a hand? I haven't got all day, ya know!"

By this time Maude and Charlie are rolling in the grass. The VW engine starts up again. Charlie hears the whine of gears and axles straining, lifts his head to see the bus going nowhere, wheels spinning, rearend lurching and writhing to clear the ditch.

Maude whispers, "Forget about it, Charlie."

Nell circles whimpering and sniffing, the pups yip and squirm on all sides, little paws to tingle Charlie's legs, groping tips of noses.

Epley races the engine, leans on the horn. Axles whine around the madly shifting gears. Charlie hears hollow metal clanking on the rocks.

"Maude, he's destroying my car."

He starts to lift his head again. She grabs his shoulders and holds him down. "It doesn't matter." Nell licks Maude's neck, tries to nuzzle in between them.

"I just put sixty dollars into that paint job," Charlie says.

"Hey! What are you guys trying to do — give me the shit end of the stick or something? Think I can pull somebody else's car out of a hole all by myself? I call that a hell of a note, buster. ONE HELL OF A NOTE!"

He bellows this, the engine roars again, yowling gears. Something rips loose with a metallic clunk and tinkle, and suddenly Epley's in the clear. The bus grinds upward through steep grass, painfully listing but pulling the grade.

"Oh Charlie," Maude begs, "don't stop now."

But he breaks their tangle of arms and legs and is galloping down the slope again, naked this time, to meet his bus as it reaches the road. Epley stops. Charlie grabs the door handle. Locked.

"Goddam you, Epley. Get outta there!"

He doesn't seem to hear. He's leaning forward slightly, like a man waiting at an intersection for the light to change, and fooling with knobs. The radio is filling, surrounding the bus:

> "Is this your first album, Eddie?"
> "It's my third, Ralph, although actually it
> isn't my album at all."
> "Can you explain that for our listeners?"
> "I wish I could, Ralph, I honestly do. Say,
> why do I keep getting little electric shocks
> from this mike stand?"

Charlie yells, shakes the door handle til his whole bus is rocking.

Epley doesn't look at him. Charlie runs around and throws open the side doors, jumps in on top of the mattress and crouches there, with palms extended, like some underweight sumo wrestler.

The black head swivels, lips apart.

"Turn off the radio, Epley."

Surprise turns to huge-eyed horror, as if Epley's struck dumb by the sudden appearance of another person in this wilderness, not just a naked man, but anyone at all.

Charlie leans and switches off the radio himself, bare shoulder next to short, black t-shirt sleeve; sweaty, Three-in-One-Oil smell of Epley, who sits now examining his knees, grabbing black thighs with grease-edged fingers.

"Get out of my car!" Charlie orders.

"No."

"Waddya mean, no?"

"What the hell do you think I mean? No. N - O."

Charlie has climbed the front seat, sitting next to him, shoving. "Out. Out."

"I call that the shit end of the stick, my friend."

"Goddam you Epley. Move!"

Epley glares at the windshield wiper. "All I want to know is, who got this car out of the ditch? Huh? Tell me that?"

"Shut up, Epley, just shut up . . ."

"Out of the ditch, singlehanded, back on the goddam road, and me with a flat tire of my own, not to mention more work at home than any ten men can . . ."

Charlie is ready to start punching, when Epley stops in mid-sentence, staring through the window, befuddled and open-mouthed. It's Maude, floating down the knollside, waist-deep in tassled grass, with Nell scampering. Under one arm she carries her puppy box, over her shoulder Charlie's paisley shirt, his corduroys.

Charlie watches too. Her knees flex slightly, smooth tendons catch the light with each downhill step. The way she walks, drifts, descends, he would recognize it anywhere now, unmistakably Maude. It fills him with pure affection.

She stops on his side of the car, holding Nell, who's growling toward Epley. Charlie pushes the window open, loves her for the wholesome barelipped smile.

"You said you were going to take me back to the highway."

Her face inches from his, her shawl loosely draped, Charlie's affection splinters into a thousand warm needles. He is weak with lust. His groin aches. He would like to run over Epley with his car.

He turns to see those startled eyes focussed on his lap, where Maude's presence has had dramatic effect. Charlie grabs the car keys, opens the door, steps out.

"Give me my pants, Maude."

Through the open window he addresses Epley.

"Listen to me."

The eyes widen.

"I'll give you a ride to the bottom of the hill."

"What good will that do?"

"What good . . . ?"

"I got a flat tire down there, buddy."

"All right, I'll take you back to your shop." .

Epley considers this, pulls the loose, scraped skin beneath his jaw.

"Can I do the driving?" he asks.

"Hell no, you can't drive! You've already destroyed my whole rear end. Spring's busted, bumper's mangled, body all mashed up. Jesus Christ, Epley!"

Charlie is starting to shout. Epley outshouts him. "Jesus Christ yourself, pal! Who hiked up from the bottom of this road? Who climbed while everybody else got a nice cozy ride? Huh? Can you tell me that?"

Nell is barking again, and Epley reaches for his scalp again, dark fingers digging into dusty, oil-black hair.

"Who sweated his ass off while everybody else is laying around in the grass taking a sunbath? Huh? I'm just asking for a fair shake. Is that too much to ask? Can you give a guy a fair shake?"

While holding Nell's taut collar with one hand, Maude is helping Charlie dress, running his belt through the loops, buttoning his shirt, tucking in his shirt tails with slow, tender hand plunges.

"Okay Epley, okay. You can drive. Straight to your shop. How does that sound? Everybody will get in back."

"Just hold on to that dog bastard. Don't expect me to drive this hill with a goddam bloodhound yipping at my neck the whole way down."

Charlie is tempted to turn Nell loose on him again. But last time that only made things worse. And already Epley's squirming in the seat like a trapped man, frantic eyes scanning, the way he looked when he shoved his lawnmower out the door.

Maude is trying to buckle Charlie's belt with one hand, fingering his navel. "Are we all getting onto the mattress?" she asks.

"Yeah. Let's load it up."

He ties Nell to the rear door, sets the puppies next to her, shoves his bike to the left, hands Epley the keys.

"There's no rush. Just take it easy. Okay?"

"Yes," Maude adds, "these curves are . . . hard on Nell."

"Easy does it, folks," Epley yells. The clutch pops. The bus bucks. He leans on the big steering wheel, nose to the window like they're heading into heavy fog. Bus bucks again, leaps forward, stuttering. Maude and Charlie are thrown back onto the mattress, shoulder to shoulder, thigh to thigh. Their bodies clamp together.

"Hey, what's going on back there?"

Rearend swings wide as the bus careens around the first curve.

"Never mind, Epley. Watch the road, for God's sake!"

Nell is still barking, sometimes gagging as her new rage strains against the rope. Pups have spilled again, and Maude is unzipping Charlie's fly. Each time Epley swings a curve the bike slides across the mattress. Charlie tries to brace it with his foot. Everything slides to the right now. On some curves the bumper and one corner of the frame scrape the ground. No time to worry about that. Maude's skirt is up around her waist, her shawl falls open.

The bus is picking up speed. With each curve the puppies tumble over Charlie and Maude. And soon she's murmuring, "Oh Charlie, oh oh oh."

At that moment Epley's tentative, quavering baritone drifts back from the front seat, nasal and phlegm-throated, feeling its way:

> *Without a song*
> *The day would never end.*
> *Without a song*
> *The road would never bend.*
> *Without a song*
> *A man aint got a friend . . .*

It grows louder, expanding as he warms to the car's acoustics, drowning Maude's little moans of ecstacy. He throws both windows open, sits back to breathe deep, and guns the engine for a short stretch of straightaway:

> *I'll never know*
> *What makes the rain to fall.*
> *I'll never know*
> *What makes the grass so tall . . .*

By the end of the first chorus Maude is covering Charlie's mouth and chin with slow, grateful kisses.

After the second chorus Charlie says, "When are you coming back from Mendocino?"

Languidly she rolls her head toward the window. With smooth-browed and transcendental smile she gazes at a stand of eucalyptus hurtling past outside. "Depends on when I get there?"

Epley, bellowing the refrain, whips the bus around a hairpin curve like it's a sportscar, slides up banked gravel into a soft shoulder, recovers, shoots down the next straightaway. Charlie closes his eyes, one foot still braced against the bike, and figures it would be riskier to try and wrest control now than to let him keep driving. He waits and prays Epley will find the brakes before they reach the ocean, and while Nell howls out a counterpoint to his triumphal song, her puppies pour back and forth over the spent lovers in a wriggling furry cascade.

The land speed record car Bluebird *completed in 1960 after five years work is powered by a 4,100 h.s.p. Bristol-Siddeley Proteus turbine engine. The car, weighing 3.37 tons is thirty feet long with a wheel base of 13 ft. 6 in. On 16 September, 1960, after reaching 360 m.p.h. in 1.7 miles the car became uncontrollable and crashed with only minor injury to Donald Campbell.*

THE GUINNESS BOOK OF RECORDS

The Parking Tower

CHARLIE BATES KNEW BETTER than to try parking along the street. The curbs were lined with cars that never moved, never left their spaces. And those cars already moving crept in endless columns, inhabited by people who had nowhere else to live or who couldn't find a parking place close enough to make it worthwhile going home. These cars were in perpetual motion, searching, circling, ready to lunge into spaces that never appeared. So Charlie had the habit of parking in a many-floored tower near his office, in the city's center.

This habit pleased him. It was the smart thing to do, the sensible adjustment to an almost intolerable situation. He was a man of the modern world, who knew what it took to survive. "You have to decide," he'd often say, "which battles you're going to fight."

One morning Charlie arrived downtown later than usual. It was nearly noon when he turned into the cool entrance tunnel and stopped at the lighted glass cage. The attendant, in white overalls, handed him half a time-punched ticket and said, "Pretty full right now, Mr. Bates. You'll just have to check each floor as you go up."

"Okay. I'll find something, Mel. Thanks."

The concrete floor was dark and shiny with the polish of a million tires. Charlie eased away from the cage's light, passed the gas pumps, aimed for the curving ramp that spiralled upward around its concrete core like a children's twisty slide. He drove confidently, sure of the territory. In his rearview mirror he saw a woman driving behind him. She looked scared, as if she feared not finding a space at all. Charlie felt superior to her. At the first landing he saw other cars searching up and down aisles for empty spaces. He smiled at their naivete.

At the second landing a quick glance told him this floor too was full. He had learned how to spot empty spaces without slowing down. He could scan an entire floor while rounding the flat place in the spiral. The woman behind him peeled off and headed down an aisle. He smiled again.

He didn't bother to look at floors three, four or five. He pitied the drivers wandering these lower levels. He took great pleasure meanwhile in setting his wheel to pull the grade without adjusting his steering. His car handled well on the incline, he thought.

At floor six he glanced right, left and straight ahead, half for the lack of anything better to do. All spaces were full. He angled into the climbing tunnel again, heading for seven. "Lucky seven," he thought. Charlie had never parked higher than seven. As the track flattened he was ready to peel off and follow an aisle to some waiting space. But his peripheral vision told him that seven too was full. Suppressing a tiny pang of terror he held the angle and entered the curving tunnel again, finding some comfort in the fact that at least he hadn't lost the set of his wheel.

Floors eight, nine, ten and eleven were full. At floor twelve he thought he saw a space far down the aisle to his right, and he peeled off, cocky now, since it wasn't a space your average driver would see from the ramp. A bulky concrete pillar nearly concealed it. At the end of this aisle Charlie turned left, ready to slip into his niche between the fenders of two other cars, only to discover there a row of closed trunks and chrome bumpers and license plates. He stopped and stared. He was sweating. It had been months since he made a mistake like that. "I'm getting jittery," he thought. "It's ridiculous."

He jammed into first gear and sped back toward the upward ramp with a violent roar, leaving a billow of exhaust fumes behind him to settle on the rows of silent cars.

He almost hit a convertible rounding the spiral track in front of him. It was the woman who had peeled off at the second level. She stopped and leaned out the window, yelling, "Hey, what's going on!"

"Sorry maam," Charlie began, "I didn't realize anyone . . ."

"Oh, *that's* all right. I mean what's going on in this parking tower? What kind of a parking tower is it when you can't find a place to park?"

"There'll be something up ahead."

"You think so?"

"Sure. I've been parking here for months, never had any trouble."

"I hope you're right. It's kind of spooky."

"Spooky?"

"Eerie, sort of. I mean, there's nobody up here, not even anybody waiting for the elevator."

"It's always like that up here. Listen, I'll tell you what. If you're worried, just fall in behind me. We'll look together. Will that make you feel better?"

"Oh thank you, yes it would, thank you very much."

She backed up. Charlie pulled in front and started to climb again.

"She's not bad looking," he thought as they passed floor thirteen. He imagined it *could* be a little scary for a pretty girl like that, up here all by herself. He set his wheel and wondered if she would notice his style on the turns.

He felt very alert now, a veteran, an old pro showing this rookie the ropes. He didn't want to be caught slowing down unless there definitely was a place to park. He passed floors fourteen, fifteen, sixteen and seventeen as if strapped to a spiralling escalator. At each level Charlie glanced right, left and straight ahead, saw the unbroken pattern of cars, then glanced in his rearview to watch the woman slow down and nearly stop while she too looked around, finally gunning her engine to catch Charlie halfway to the next floor. As levels flashed past them, in his mirror he could see the woman's face contorting strangely. At floor twenty she began honking her horn and waving. Charlie stopped at the next landing and walked back to her door.

She yelled, "What in hell is going on around here! What in hell is going on!"

Charlie was dismayed. "I really don't know, maam, I wish I did."

"Well, I don't know about you, but I'm going back down! I can't take any more of this!"

He watched her face, framed in the open window, and felt the first warm rush of desire. He nearly leaned forward to kiss her mouth. "I'd agree with you about that, except I really don't know what good it would do to go back down. There just *weren't* any parking places, you saw that. Get back out in the street, you're no better off."

"What are we going to do then?"

She shouted this as if it were all Charlie's fault. Seeing his puzzled look, her face collapsed in tears. Her head fell forward against the arm that lay along the window sill.

"Now you just relax for a minute, maam, and we'll start climbing again."

"Oh I can't. I simply *can't!*"

"We have to, don't you see? We don't really have much choice in the matter."

She was sobbing. Filled with compassion he reached down to touch her arm. She grabbed his hand.
it with frantic kisses.

"Don't leave me here. Please don't leave me here alone."

"Nobody's leaving you here. C'mon, snap out of it. Few more floors, we're bound to find a spot. C'mon. Let go of my hand and sit up."

She pushed his sleeve back and started stroking his wrist.

"Hey, cut that out! Get hold of yourself." Charlie jerked his hand away, and she sat up abruptly, wiping her eyes.

"I'm sorry. I'm being very foolish. You're absolutely right."

He gave her his handkerchief. She blew her nose. He was about to return to his own car when she reached for his hand again, clutched it tightly, and said, "But I like you, you know. What's your name?"

"Bates. Charlie Bates. What's yours?"

"Deedee."

"Deedee?"

"Ummhmm."

Giving in to his impulse Charlie leaned forward and kissed her passionately. Deedee opened her lips to him.

At that moment another car rounded the track from the floor below and nearly rammed Deedee's rear bumper. The car stopped and a florid man stuck his bald head out the window.

"Hey, what the hell's going on around here!"

Charlie straightened up, wiped his mouth with his hand. "Sorry mister. Just trying to help this lady . . ."

"The hell with her. Where did all these goddam cars come from?"

"That's what *we're* trying to figure out," shrugging, smiling.

"Good Christ, I was supposed to be at a meeting half an hour ago, and I've spent the whole goddam time on this fucking merry-goround."

"It's aggravating, I'll grant you that."

"Well then get moving, for Christ sake!" The man leaned on his horn, filling the cavern of floor twenty-one with a hollow blare.

Charlie jumped into his car and headed upward again, followed by Deedee and the florid man, who kept blaring his horn. The noise unnerved Charlie. He found his speed increasing steadily with every upward turn. By the time they reached floor thirty he was going forty miles an hour. Deedee kept up with him, and the man's horn seemed to grow louder the farther he got from his meeting.

At floor thirty-nine the man peeled off. Charlie knew there was no room. In his rearview he could see the scarlet face, it looked ready to burst. He could hear the constant horn echoing over roofs of the thousand cars packed tightly there, the frantic cry of a crazed animal in alien surroundings. He heard gears grinding and screaming as the wild car snorted among the aisles, and from the floor above, Charlie heard the horn stop and the final

50

reverberating crash as the florid man tried to force his way into the narrow space between two parallel sports cars.

Charlie's wheel was set, and the speed was hypnotizing him. He was going forty-five. He noticed they were passing floor forty-five, and he thought he saw some meaning in that. He tried to increase his speed a mile per floor, to see how fast they could actually climb. His car was performing beautifully, he thought. He glanced at the temperature gauge. It had risen some, but there was nothing to worry about.

At floor fifty-five Deedee began honking again, waving with one arm. "She shouldn't do that," Charlie told himself. Then, as if he'd prophesied it, her right fender caught the wall when she flattened for fifty-six, gouging out a hunk of concrete. He knew she was scared and losing her nerve, but he kept climbing. There was no alternative. Every floor was full.

He was going an even sixty when they passed that floor. Behind him Deedee's horn blared erratically while she careened from wall to wall in the winding corridor. She was out of control and Charlie regretted it. He couldn't stop now, though. At floor sixty-one, in desperation Deedee peeled off for the last time, still honking out some strange telegraphy, and Charlie heard her crash into the row of cars nearest the ramp, heard the metallic explosion, and no screech.

"She didn't even use her brakes," he thought. He shuddered and shook his head, and, although he'd reached floor sixty-three, he cut back to sixty miles an hour. He could make the curve easily at sixty, as long as his wheel held and the radiator didn't boil.

He was proud of his car, a little proud of himself too, going it alone now, without the comfort of Deedee behind him. Amazing, among other things, how his perception had improved. Even passing a landing at sixty he could tell in an instant whether or not there was space to park, though in truth he no longer paid much attention to that. At this point he didn't really believe there'd be a parking space, and somehow it didn't seem to make much difference. The parked cars dropped past him in free-falling layers. He began to lose count of the floors. The continual upward spiral made him dizzy. The floors blurred, and for a while he couldn't tell which way he was headed—up, down or sideways. Then miraculously his vision cleared, the dizziness passed, replaced by a marvelous giddy well-being, a light-headedness that made him feel all-powerful, convinced him he could do no wrong, make no mistakes. Was it lack of oxygen, he wondered, a rarifying atmosphere? He knew such a thought should frighten him. But it didn't.

It didn't matter. Nothing mattered, nothing could go wrong because now he could make no mistakes.

Charlie sped upward, whirling and whirling, past floor upon floor of the parking tower, until at last he saw a radiance ahead. The ramp leveled out, and he broke into sudden, eye-piercing sunlight. The road widened, as other tracks like his fed into what became a twelve-lane freeway crossing what appeared to be an endless plain. In five of the other lanes he saw drivers like himself, blinking in the bright light and looking around. They all sped along at sixty, abreast now, and they started waving to each other.

Charlie yelled to the man nearest him, "Hey, we made it!"

The man yelled back, "Yeah, how about that?"

Anxious to share his elation Charlie shouted, "Some road, huh!"

"You betcha!" the man replied.

The other men were yelling at each other, and they raced together like that for miles, until the man next to Charlie shouted in another tone, "Hey, what's that up ahead?"

Charlie squinted, saw what appeared to be a row of cars coming from the other direction. He yelled, "Jesus, I don't know!"

The distant spots grew bigger, approaching at high speed. The other man yelled, "Hey, am I seeing things? Or are they in our lanes?"

"That's what it looks like," Charlie shouted, "but that's impossible!"

He leaned toward his windshield, peering through the glare. Another car was headed right for him. He glanced at his speedometer. Eighty miles an hour. For the first time he noticed that little mesh fences about as high as his windowsill separated him from the other lanes, so there was no changing lanes without crashing into a fence. He jabbed his brake. His tires grabbed and squealed, his rearend spun, tearing at one of the fences. When he finally stopped, his tail hung over into the adjoining lane. His companions kept going, and while he sat there he watched five simultaneous explosions on the road ahead, as ten cars collided head-on at eighty miles an hour, hoods, trunks, fenders flying in all directions, the hulks aflame and burning fiercely.

A moment later the car that should have hit his came barreling past, narrowly missing Charlie's front end by swerving in toward the opposite fence.

Charlie yelled, "Hey!"

The man turned, surprised. His car angled into the fence, crossed the next lane, went through another fence, through six fences that way, until it hit an anchored fence post and exploded.

Shaken, Charlie started his car. His doors and fenders were badly smashed, but the engine and wheels seemed okay. Cautiously he proceeded forward, squinting in the bright sunlight, trying to see down the long lane ahead. No more cars appeared. After several miles the road began gradually to dip. The freeway divided into several strands. He was travelling a single track again, and before he realized what was happening he had passed a sign that said DOWN RAMP.

The grade inclined sharply, curving toward a murky tunnel, and his car was picking up speed. Charlie jabbed for his brakes. But he had torn something at the fence, and they wouldn't hold.

⋖⋗●⋖⋗

Thirty spokes
Share one hub.
Adapt the nothing therein to the purpose
in hand, and you will have the use of
the cart.

LAO TZU

⋖⋗●⋖⋗

GAS MASK

I

CHARLIE BATES DIDN'T MIND the freeways much. As he often told his wife when he arrived home from work, he could take them or leave them alone. He listed freeways among those curious obstacle-conveniences with which the world seemed so unavoidably cluttered. Charlie was neither surprised nor dismayed, then, when one summer afternoon about five-thirty the eight lanes of traffic around him slowed to a creep and finally to a standstill.

He grew uneasy only when movement resumed half an hour later. His engine was off; the car was in gear; yet it moved forward slowly, as if another car were pushing. Charlie turned around, but the driver behind was turned too, and the driver beyond him. All the drivers in all the lanes were turned to see who was pushing. Charlie heard his license plate crinkle. He opened his door and stood on the sill.

He was on a high, curving overpass that looked down on a lower overpass and farther down onto a twelve lane straightaway leading to the city's center. As far as Charlie could see in any direction cars were jammed end to end, lane to lane, and nothing moved. The pushing had stopped. Evidently there was nowhere else to push.

He looked into the cars near him. The drivers leaned a little with the curve's sloping bank. Nobody seemed disturbed. They waited quietly. All the engines were off now. Below him the lower levels waited too — thousands of cars and not a sound, no horns, no one yelling. The silence bothered Charlie, frightened him, until he admitted this was really the only civilized way to behave. "No use getting worked up," he thought. He climbed back in and closed the door as softly as he could.

As he got used to the silence he found it actually restful. Another half hour passed. Then a helicopter flew over, and a loudspeaker announced, *May I have your attention, please. You are part of a city-wide traffic deadlock. It will take at least twenty-four hours to clear. You have the choice of remaining overnight or leaving your car on the freeway. The city will provide police protection through the crisis.*

The copter boomed its message every fifty yards. A heavy murmur followed it down the freeway. The driver next to Charlie leaned out his window. "Are they nuts?"

Charlie looked at him.

"They must be nuts. Twenty-four hours to clear a traffic jam."

Charlie shook his head, sharing the man's bafflement.

"Probably a pile-up farther down," the man said. "I've seen em before. Never takes over an hour or two. I don't know about you, but I'm sticking it out. If they think I'm gonna leave my goddam Mustang out here on the freeway, they're all wet."

His name was Arvin Bainbridge. While two more hours passed, he and Charlie chatted about traffic and the world. It was getting dark when Charlie decided he at least should call his wife. Arvin thought the jam would break any minute, so Charlie waited a little longer. Nothing happened.

Finally Charlie climbed out, to look for a phone booth. The nearest exit was two miles away. Luckily Arvin had a tow rope in his trunk. Charlie tied it to the railing, waved his thanks, swung over the side and hand-over-handed to the second level. From there he slid out onto a high tree limb and shinnied to the ground.

Gazing up at the freeway's massive concrete underside and at Arvin's rope dangling far above him, Charlie knew he'd never climb back. "What the hell," he said to himself, "I might as well go home. The cops'll be around to watch things. And the car's all paid for."

He began searching for a bus or a cab. But everything, it seemed, was tied up in the jam. In a bar where he stopped for a beer to cool off, he learned that every exit, every approach, every lane in the city's complex freeway system was plugged. "And ya know, it's funny," the bartender told him, "there wasn't a single accident. It all happened so gradual, they say. Things slowed down little by little, and the whole town stopped just about at once. Some guys didn't even use their brakes. Just went from one mile an hour to a dead stop."

It took Charlie two hours to walk home. His wife, Fay, was frantic.

"Why didn't you call?"

"I started to, honey."

"And what happened to your pants?"

He glanced uncertainly at his torn slacks. "I was shinnying down this tree. Maybe there was a nail."

"For God's sake, Charlie, this is no time to kid."

"I'm not kidding. You're lucky I got down at all. Some of the guys are still up there, the older guys, the fat ones, couldn't get over the rails. And a lot of guys wouldn't leave. Probably be out all night."

She looked ready to cry, and she stared as if he were insane. "Charlie, please . . ." He put an arm around her and drew her close. "What happened, Charlie, where have you been?"

He guided her to the sofa. When they sat down, his hairy knee stuck up through the split cloth. "I thought you'd see it on TV."

"See *what* on TV?"

He told her the story. By the time he finished she was sitting up straight and glaring.

"Charlie Bates, do you mean you just left our car on the freeway?"

"What else could I do? I couldn't stay out there all night. Not in a Volkswagen. I'd catch cold. I'd be all cramped up."

"You could've moved into somebody else's car. This Arvin fellow would have let you. Somebody with a good heater or a big back seat."

"You can't just barge into somebody else's car and stay overnight, Fay. Anyhow, I wanted to phone. That's why I came down in the first place."

She rubbed his bare knee. "Oh, Charlie." Leaning against him again she said, "At least nothing happened to *you*."

She snuggled there and they were quiet until she said, "But Charlie, what'll we do?"

"About what?"

"About the car?"

"Wait it out, I guess. Wait til tomorrow at least, until they break the jam. Then get back over there. Of course, that won't be as easy as it sounds. Probably have to get over to the nearest ramp and hike it — maybe two, three miles of freeway, up the center strip, I suppose — plus getting to the ramp itself, which is right in the middle of town. Maybe I can borrow a bike. I don't know."

"Don and Louise have a two-seater. We could borrow theirs and both go."

"Maybe," Charlie sighed. "Let's worry about that tomorrow. I'm bushed."

Next morning Charlie borrowed the big two-seater from Don and Louise. Fay packed a lunch, and they pedalled across the city, figuring to get there early, to be on hand when the car was free, although an early solution was no longer likely. The news predicted another thirty-six hours before traffic would be moving. The jam now included not only the freeways, but all main streets and intersections, where busses, trolleys and trucks were still entangled. It even extended beyond the city. Police had tried to block incoming traffic, but that was impossible. All highways transversed the city or its net of suburbs. Impatient motorists, discrediting reports, finally broke the road blocks, and the confusion was extending in all directions by thousands of cars an hour.

Smugly by-passing all that, Charlie and Fay followed an ingenious route of unblocked streets he mapped out after watching the news. Near the freeway they mounted a high bluff and decided to ride an elevator to the roof of an apartment that rose above the overpass where their car was parked. Charlie brought along a pair of navy binoculars. From that vantage point they ate lunch and surveyed the curving rows of silent cars.

"Can you see ours, Charlie?"

"Yeah. She looks okay. A little squeezed up, but okay."

"Lemme see."

"Here."

"Some of those poor men are still sitting out there." Fay said. "Don't you know their wives are worried."

"Their wives probably heard the news. Everybody must know by now."

"Still worried, I'll bet." She hugged Charlie and pecked his cheek. "I'm glad you came home." Then, peering again, "I'll bet those men are hungry. Maybe we should take them some sandwiches."

"Take a lot of sandwiches to feed everybody stuck on the freeway."

"I mean, for the men right around your car. That Arvin, for instance. You know . . . your neighbors."

"I don't really know them that well."

"We ought to do *something*."

"Red Cross is probably out," Charlie said. "Isn't that a cross on that helicopter down there by the city hall? Here, give me the glasses."

"It is." Fay said. "They're dropping little packages."

"Here. Let's see. Yeah. Yeah, that's just what they're doing. Guys are standing on the roofs of their cars, waving."

"The poor dears, they've had a rough night."

"The poor dears."

Charlie munched a tuna sandwich and scanned the city like a skipper. After a few minutes Fay pointed, "Look, Charlie. Over that way. Two more helicopters."

"Where? Down there? Oh yeah. Couple of military birds. I guess the army's out too."

"What're they doing — lifting one of the cars?"

"No, not a car. It looks like a long, narrow crate. And they're not lifting it, they're lowering it endways. A couple of guys in overalls are down below waiting for it. There. It's down. They're anchoring it to the center strip. Wait a minute. It's not a crate. One of the guys in overalls just opened a door, and he's stepping inside. Hey. People are jumping out of their cars and running down the center strip. They're running from everywhere, climbing over hoods. Somebody just knocked over the other guy in overalls. I think there's going to be a fight. They're really crowding around that door and pushing. No. I think it'll be okay. The guy inside just came out. He's tacking a sign over the door. All the men are starting to walk away. Women are lining up along the center strip."

"Wouldn't you know it."

"A woman just opened the door and stepped inside."

"What if the cars start moving while she's in there, Charlie? That's what I'd be worried about."

"It's a risk, all right, one of those risks a person just has to take."

III

From the rooftop they could hear the police copter's periodic messages. By the end of the first day, predictions for clearing the jam were at least two, perhaps three more days. Knowing they should be on hand whenever it broke, yet weary at the very thought of pedalling across the city twice each day to their vantage point and home again, they decided to rent an apartment in the building below them. One happened to be available on the top floor, facing the freeway. They moved in that evening, although they had little to move but binoculars and a thermos. They agreed that Charlie would pedal home the next day and pick up a few necessities, while Fay kept an eye on the car.

The plan worked marvelously. Once situated they set up a rota-

61

tion watch, four hours on, four hours off. Charlie figured he could reach the car in half an hour if things looked ready to break. He figured he'd have that much warning, by listening to police reports and watching TV and checking the progress downtown where the cranes worked. Through the binoculars he watched great jaws lift out cars, vans and busses and drop them over the sides of the freeway. Things would loosen up down there first, giving him time to bicycle six blocks to a pine tree a mile below his car. Scaling the tree he could reach the top of a fifteen foot concrete retaining wall and drop to the asphalt. From there it was an easy jog up the center strip and around the sloping clover-leaf curve to the over-pass.

To be safe Charlie made dry runs over this course a few times each day. He'd switch on his engine and warm it a while, then stroll back, greeting the stranded motorists, who watched his passage with mixed admiration, envy and disbelief. By the third day the men were stubble-faced, sullen, dark-eyed from fitful sleeping. The women were disshelved, wan, most of them staring blankly through windshields at nothing. Charlie felt he ought to do something. Sometimes he squatted on the center strip to talk to the man who'd lent him the tow rope.

"How's it going, Arv?"

"Bout the same, Charlie."

"Pretty hot out here today."

"Bout like it's been, Charlie. Getting used to it, I guess. You probably feel it more than I do. That's a long pull."

"Not so bad anymore. The old legs are shaping up."

"How's your time?"

"Twenty-eight, ten, today."

"Cuttin it down, hey boy."

"Poco a poco," Charlie said. "Poco a poco. It's the elevator that holds me back. Slowest elevator I've ever seen."

"You thought of waiting on the sidewalk someplace? The wife could maybe signal out the window when the time comes."

"Say, Arv . . ."

"It came to me yesterday, but I figured you'd already thought of it."

"Never entered my head. That's a great idea."

Charlie paused. "I've been meaning to ask you, Arv," he went on. "Why don't you come up to the apartment to meet Fay. I've told her about you. You'd like her, I know. We could have a couple of drinks, relax awhile."

"That's real nice of you Charlie. But . . . I'm not sure. The

62

trouble is, you never know when the thing's going to break loose."

"I've got that two-seater. If anything happens we can pedal back over here in no time. Cuttin' it down every trip, ya know. C'mon. It'd be good for you to get away."

"I'd like to Charlie, I really would. But to be honest . . . I haven't had this car very long. I'm still making payments, and . . . well, I just feel like I ought to stick pretty close to it."

"I know how you feel, Arv. In a way I don't blame you. I get a little jumpy myself, especially at night when I can't see much. But look, if you change your mind, I'll be back this afternoon."

"Thanks, Charlie."

"See ya later, Arv. And thanks for that idea."

"My pleasure, Charlie. Hate to have you miss your car when the action starts."

IV

Taking Arvin's advice Charlie spent most of each day sitting on a bus-stop bench across the street from the apartment house.

At last, on the afternoon of the sixth day after Traffic stopped, Fay's white handkerchief appeared in the window. Charlie's bike stood before him in the gutter. He mounted it over the back wheel, like a pony express rider. In a moment he was off and pedalling hard for the pine tree.

From blocks away he could hear the now-unfamiliar rumble of a thousand engines. As he gained the top of the concrete wall and poised ready to drop, a cloud of exhaust smoke swirled up and blinded him. It stung his eyes. He began to cough. He dropped anyway, sure of the route he must follow, even if he couldn't see. Gasping and wiping his eyes he clambered over hoods toward the center strip. The smoke didn't abate. It puffed and spurted, choking Charlie. Every driver was gunning his engine, warming for take-off. In a panic that he'd miss his car, that it would be carried away in the advancing stream, Charlie stumbled blindly upward, deafened by the sputtering thunder of long-cold cylinders, nauseated by fumes, confused by the semi-darkness of gray, encompassing billows.

The cars disappeared. It seemed he staggered through the smoke for hours. He nearly forgot why he was there, until he heard a yell behind him. "Hey Charlie! Where ya goin?"

"That you, Arv?"

"You just passed your car!"

"This damn smoke."

"Helluva thing, isn't it?"

Arv was elated. Through the veil of fumes that curled up from under Arvin's car, Charlie could see a wild expectancy lighting the haggard eyes. His yellowed teeth grinned behind the beard.

"What's happening?" Charlie gasped, hanging onto Arvin's aerial while his lungs convulsed.

"Looks like we're moving on out. Better warm up."

"When did you get the signal?"

"No real signal," Arvin shouted, "but everybody down the line started up, so I started up. Things ought to get going anytime."

"Have you moved at all?"

"Not yet, but you better get the old engine cooking, Charlie. We're on our way, boy! We're on our way!"

Coughing and crying Charlie staggered to his car, climbed in and switched it on, accelerated a few times, then leaned forward to rest his head on the steering wheel. Nausea overcame him. Engine roar throbbed against his temples. He passed out.

When he came to he was staring through the wheel at his gas gauge: empty. He looked around. The smoke had cleared a little. In the next lane he could see cars vaguely outlined. None had moved. He switched his stuttering engine off. Others were doing the same. The rumble diminished from moment to moment. Within minutes after he came to, it was quiet again. There was no wind. The smoke thinned very slowly. Behind him he saw a driver sprawled across the hood, chest heaving. In front of him a man and woman were leaning glassy-eyed against their car. In the next lane he heard the wheeze and rattle of a man retching. He turned and saw Arvin leaning out his open door.

The police copter droned toward them, hovered, sucking up smoke. *Please turn off your engines. Please turn off your engines. The deadlock will not be cleared for at least another thirty-six hours. You will be alerted well in advance. Please turn off your engines.*

No one seemed to listen. The copter passed on. Charlie climbed out, still queasy but able to stand. Arvin sat on the edge of his seat now, bent forward, head in his hands.

"You okay?" Charlie stood looking down for several moments before the answer came.

"I guess so."

"False alarm, huh?"

Arv grunted.

"Looks like tomorrow could be the day, though," Charlie said.

Arv nodded, then raised his head. His eyes were dark, weary, defeated. All hope had left him. Deep creases of fatigue lined his cheeks and forehead. His beard was scraggly and unkempt. He looked terribly old. With a hoarse and feeble voice he said, "But Charlie . . . what if it's not tomorrow? What're we gonna do, for God's sake? It's been six days."

"Look, Arv. You heard the announcement. Another thirty-six hours at least. Why don't you come on up to the place and lie down for a while?"

A little light brightened Arvin's eyes. His mouth turned faintly toward a smile, as if remembering some long-gone pleasure. But he said, "I can't, Charlie." He raised his shoulders helplessly. "I just can't."

Charlie nodded. "I know, Arv. I know." After a pause he said, "I guess I'll see you in the morning then." He waited for Arvin's reply, but the head had fallen again into the palms of his hands, and Arv sat there swaying. Charlie walked away.

Most of the smoke had cleared. The heavy silence was broken now by distant groans, staccato coughs. All around him, down the curve he would walk, on the other freeways that snaked so gracefully below him, in among the rows of dusty cars, he saw people sprawled, hunched, prone on the center strip, folded over fenders, hanging out windows, wheezing, staring, stunned.

He picked his way to the concrete wall, scaled it and left the devastation behind. He knew, though, he'd have to return, perhaps several times. No one could tell when it would be over. The police reports were meaningless. He returned to the apartment to console Fay, who felt guilty about sending him on a wild goose chase. Then he pedalled downtown to a war surplus store. His lungs still burned from the smoke. He decided to buy Arvin a gas mask and one for himself.

The Wisdom of Y.T. Belinsky

Spinning Things

Watch the hubcaps, Charlie thinks, the myriad shapes they take, the rings, bowls, plates, cups, flames, blades, spokes. Some, spinning, seem fixed. Others are designed to exploit the spin, their curves and angles catching light. The spokes flash. Flames dart out from silver fires. Insignia stamped on the metal become small dark circles. These things beguile Charlie. He thinks of his own hub and wheel and black squash of elephant flesh resting on the concrete pad back there where it settled during the night. The final insult. Now, at sunrise, he has reached this spot at the edge of a culvert where he sits in shrubbery and watches the spinning hubs, and if he could watch long enough something surely would be revealed to him. But after five minutes he has to turn his eyes from all that spinning. A little queasiness creeping in.

Submarine Novelties

Out of the culvert, standing on the gravel shoulder, he tries to see the faces passing. With quick glances drivers size him up, suspicious of a man alone by the roadside at this hour of the morning. Smug whizzers. Arrogant bastards, Charlie thinks. He knows the feeling well. Inside a car you feel superior to those on foot. Approaching a hitchhiker, you hold all the cards, can decide to stop or not stop, while he has to stand out there awaiting your decision. At the service station you watch a man in a windbreaker lean toward you with his spray bottle. Being outside your car, beyond your curve of glass makes him somehow vulnerable. He is trained not to look at you, yet his face is so close you can count the nasal hairs. You are in a diving bell, he is the submarine novelty swimming past.

Who is Charlie?

In the eyes of passing motorists a submarine novelty now himself.

A man who refuses to fix his own flat tire.

A man in the middle of life, with blue eyes, a blue workshirt, brand-new logger boots, and forty-eight dollars in his wallet, having been deeply affected by a series of ads for travellers checks that ran for several months when he was in his twenties, showing famous personalities like Bob Hope declaring that they never carried more than fifty dollars in cash.

A man who turns to contemplate his own vehicle with, for the first time in his life, total detachment. Free of anxiety, free of affection, free of hatred, pride, despair, fury or awe, he sees it clearly for the monstrosity it is. Some chance alignment of tree and hillside surrounds it with the new sun's orange rays. He thinks of the Boy Scout knife he owned as a kid, the kind with blades for every imaginable purpose, from threading needles to felling small trees. The trick was to unfold the knife so all its features showed at once — the blades and spoons, canopeners, screwdrivers and fish-gutters all sticking out at twenty different angles. That is how Charlie's car looks to him now. Not a car at all any more. His wife talked him into trading both their cars for this red and silver camper truck, with aluminum siding that glitters like tinfoil, a translucent water can strapped to the front fender, spotlights beside each window wing, a shortwave aerial, atop the cab two scubadiving tanks, an orange plastic dinghy on the roof, with yellow oars, next to a square silver air vent like a small cupola, and up each side of the rear door two Japanese motorcycles, one for him, one for Fay. "So we don't feel like we're chained to the camper," is the way she put it.

A Malignant Tumor

Connecting the cab to the kitchen-bed-sittingroom is an intercom with two-way control, and this is what finally got to Charlie, about an hour after they set out, yesterday morning. Faye, tired from packing, said she'd lie down in back for the first couple of hours and trade off on the driving later. They were heading for this State Park about a hundred miles away, Charlie flashing forward to the campsite, dinner under redwoods, and then the secret pleasure of pulling all the shades and making love in their portable motel. He was thinking of an article he'd read on how a change of scenery, sometimes even a change of rooms, can be an aphrodisiac, when the intercom clicked on. Like a gravelly message from the control tower, Faye's thin filtered voice said, "Charlie? Charlie, can you hear me?"

He lifted his hand-mike from the dashboard and leaned into it, smirking. "No, I can't hear a word you're saying."

"C'mon Charlie, just tell me yes or no."

"Yes or no. What's the matter?"

"Nothing. I'm just not as tired as I thought I was."

68

"You want to sit up front?"

"Not right now. I kind of like it back here."

He turned around, saw her face about twelve inches from his, beyond two panes of glass. Something in the glass itself, the cage it made, irked him, extinguished his lust. She was lying on her stomach, with the mike to her lips, a playful kitten whose brow bunched suddenly in terror.

"Charlie! Look out! There's a truck! Coming around that curve!"

At her cry he spun the wheel, swerved over the center line, saw then that the truck was still a hundred yards away. As he straightened out, cursing, he switched off the speaker. Faye switched it back on.

Long-distance caller. "Charlie?"

"What?"

"Did you switch me off?"

"Yes."

A long pause, then Faye said, "Is that the way it's going to be, Charlie?"

"What do you mean, is that the way it's going to be?"

A longer pause. Charlie turned again, saw her watching him through the double pane, holding the mike away from her face, loose-wristed, as if it were the handle of a very long cigarette holder. It became one of those small moments where twelve years of married life needle down to a pure point of complete silence, every word, every question and answer by this time so familiar and inevitable one look contains whole hours of heated, circular talk. Silence, and this particular struggle is reduced to the matter of who will switch off the intercom, it being an act of generosity to leave open their narrow line of communication, a petty and hostile one to abruptly close it.

Pointless static prickled from the speaker, and as it filled the cab Charlie saw with sudden clarity, and no surprise, as if he had known it all along, that he had reached his limit, this intercom the image of where he and Faye had arrived. It would only be a matter of time now, of deciding when to make his move. What had Faye done to provoke this? Nothing you could put your finger on. It was just that the burdens of his life, it seemed, were all domestic, taking the shape of 1.) property, and 2.) equipment. To be sure, they had spawned this encumbrance together. But their relationship was by now so entangled with all they'd collected, whatever sentiment they'd begun with twelve years back had strangled somewhere under the cars and lawns and waffle-irons and scuba-tanks and dryers. And this camper rig, which cost Charlie over five thousand dollars to get rolling, he saw as some grotesque and final tumor swelling up from all the rest. He has to admit it wasn't entirely Faye's idea. He just forced her to talk him into buying it. But that too is typical of the things they spend their time doing these days. This tumor, Charlie fears, is one that can't be removed without killing the patient.

How Does Charlie Feel?

Regretful? Yes. Sorrowful? No. Foolish? A little. But now there's no question in his mind. Five thousand dollars worth of brand-new equipment and you wake up to a flat tire. Faye is still asleep inside the tilted room, her legs following the mattress slope. She'll be all right. Everything he has is hers. A paid-for house to live in. Money in the bank. The camper, the intercom, the dinghy, the motorcycles. A thirty-four year old woman with two motorcycles should be able to attract a lover without much trouble. All Charlie wants, as he tells the driver of the station wagon that pulls off the highway, is space and time.

"What do you want?" the man yells over a clattering, cam-ruined engine.

"Space and time," Charlie shouts.

"I saw you waving your arms!"

"No you didn't!"

"I didn't?"

"I was swinging my arms. There's a difference."

"You want a ride?"

"Sure," Charlie says and climbs in.

"Space and time, eh," the man grins, flashing teeth through his beard.

Charlie nods, instantly afloat with the hitchiker's special pleasure, having it both ways: off and running, with an early start, yet his forward motion rests in other hands. Static, oars, aerials, power mowers fall from his shoulders like dry leaves.

"The name's Y.T. Belinsky."

"Charlie Bates."

"How old are you, man?"

"Forty-seven," Charlie lies.

"You don't look it."

"Thank you."

"What are you going to do?"

"Don't know yet."

With eyes both languid and intense, Y.T. regards him, grins

70

again, leans into his steering wheel as if something down the road amuses him, and Charlie gets the impression that Y.T. too is on the loose, perhaps for the first time, and just getting the feel of it. Intuitively Charlie trusts him. He leans back against the spongey seat, lets his head drop, stretching the neck, and sees from this position that the ceiling upholstery has been torn away and a sign is painted on the rough metal underside of Y.T.'s roof. It says,

THE WISE TRAVELLER LEAVES NO TRACKS.

Charlie knows Y.T. knows he is looking at the sign, and he knows Y.T. won't ever mention it.

"Listen to this," Y.T. says. He turns on the radio, randomly fooling with the dial til the static clears. They hear a high pitched hysterical laugh, deep moans in the background, and cries of anguish, then a creaking door and a sinister voice announcing the grand opening of *The Laugh In The Dark Car Wash*. Unctuously it says,

> That laughter you hear is just another terrifiied customer. They all end up that way. Deranged with horror. If you don't drive out of here screaming we'll wash your car for nothing. In fact, if you can get here between 10 and 12 today, we'll wash it for nothing anyway. Come on in. The water is fine.

Y.T. switches off the radio, throws Charlie a sidelong, seductive look and says, "You wanna try it, man? We can make it by the time they open."

Ready for anything, Charlie says, "Why not?"

Charlie's First Lesson

Two benevolent monsters guard the entrance to *The Laugh In The Dark Car Wash*, grinning down at cars that pass between them. One resembles the Hunchback of Notre Dame, the other a tyrannosaurus. A rubbery black fringe hangs almost to the roofs of the cars. Y.T. passes under this, turns his engine off as the conveyor belt picks them up.

"Scoot toward the center, man. That window wing is busted."

Water assaults the car from ten directions, as if they are trying to drive through the Panama Canal at night and locks are filling up all around them. From out of the darkness metal arms with mops for hands start pawing at the doors. Phosphorescent white lines make them look like arms of skeletons.

"One thing we all learn," Y.T. says, "is care and respect for the vehicle. Right, Charlie?"

71

"That's right." Charlie thinks of all the years he has squandered at that very occupation, and how liberated he feels coasting along with Y.T.

"I mean, you wash it, you sweep it out, you cover it at night, you water it, oil it, gas it, lubricate it. If it breaks down, you drop everything else you're doing to get it to the garage. You will actually go in debt to keep it running. Isn't that true?"

"You got anything I can throw over this window wing, Y.T.?"

The water has started again. This time they are driving under Niagara Falls. A rusty place in the roof has given way, and water splashes down on both of them, plastering hair to their foreheads.

"The great irony," Y.T. continues, "is that we have not yet applied anything like this kind of care and respect to our true vehicle."

"Which is?"

"Why, the one we've been riding all the time, man. The planet. The earth. This spinning globe. Let's take care of that vehicle, for Christ sake! I mean, some dude will spend three thousand dollars trying to get a car that will do forty miles an hour faster than the one he has, and he is already doing sixty-five thousand miles an hour, all day long, every day of his life. How can anybody in Detroit or Cape Canaveral compete with *that?*"

Charlie stares at Y.T. and starts to giggle, at the dripping face, and at the incredible simplicity of this speech. It is a revelation. Goosebumps prickle Charlie's arms.

From the ceiling green furry limbs reach down, with soapy mittens that swirl across the windows. Hideous laughter cackles from loud speakers somewhere overhead. Bizarrely shaped brushes jab at the car, snag and tear on jagged metal edges. One brush lunges through the broken window wing, as if to take Charlie by the throat. It can't get out and breaks off, dangling there, dripping soap onto the split upholstery. This jams the carwash machinery, the conveyor belt stops. Lights go on behind a thick pane of glass, like a TV control room, where a livid man is jumping to his feet, pushing at buttons and levers. Through the loudspeaker they hear his voice. "What the hell!"

Y.T. turns his engine on, rams into low gear and starts for the exit about fifteen yards ahead. His tires bounce on the rollers and rachet wheels that keep the belt moving. Human hands dart out like the ghostly mops and brushes did, trying to stop him. Through the speaker the furious voice pursues him.

"Hey! Goddam it! Jesus Christ!"

The station wagon breaks into sunlight. Taking a shortcut over

72

the curb, Y.T. leaps between a pair of tombstones and swerves into the street, careening out of town, with soap streaming back along the chevvy's faded green like veins of foam across a rising wave.

Getting Somewhere

Y.T. is twenty seven, with a yellow stubby beard rounded at the corners. His hair is yellow too, lying back and bunched behind. His eyes, though hazel, are flecked with yellow and catch and send out all the color from his yellow dashboard. At one point, when their eyes briefly meet, in amusement over this getaway, Charlie is reminded of a zoo lion he once tried to stare down. Inside her cage, the lion's great yellow eyes seemed filled with every emotion from tenderness to hate, and Charlie was on the verge of speaking when the lion tipped her head and roared from the chest a deep carnal thunder and began to pace back and forth behind the bars.

Below his hair Belinsky wears a dirty white scarf, and a rust colored Harris tweed sportcoat, and white bermuda shorts, hoping to resemble, as he now tells Charlie, an RAF pilot on the run.

"On the run?"

"Deserting, man. Leaving it all behind."

"Why an RAF pilot?"

"Oh, you know. The dash. The verve. The bravado."

Y.T. tosses back one end of his greasy scarf, squints into the windshield as if Messerschmidts approach.

"What are you leaving behind?"

Y.T.'s answer has a strange effect on Charlie. It titillates him in an almost sensual way, as if, while taking the second long swallow from a glass of red wine, the first already tingles warmly in the blood.

"The nineteenth century," Y.T. says, "and most of the twentieth."

Charlie chuckles nervously, expectantly.

"Nothing adds up anymore. Isn't that so, Charlie?"

"Absolutely."

"And why *is* that?"

"I'd like to find out."

"It's because we are all out of phase, man."

"With what?"

"Why, with the world, man, the globe."

"How do you mean?"

"Did we say it was spinning?" Y.T. asks.

"Like a sonofabitch."

"But we keep trying to go in straight lines. Right? I mean,

look at this fucking highway we're on. Like a string, all the way to Los Angeles. Look at your life. Like a corridor, man. Until this morning you thought you were getting somewhere. Isn't that so?"

Y.T. turns with an odd smile, a saintly smile is the way it strikes Charlie, the eyes filled with both a liquidy benevolence and some terrible zeal. Again Charlie feels the goosebumps, is made aware of every inch of skin beneath his clothes, that strange wine flowing. Notions percolating inside him for months suddenly sparkle clear. In his mind's eye brilliant hubcaps whirl out of sight in all directions.

"Yet those very words give it away," Y.T. goes on. *"Getting somewhere.* It is the Laugh in The Dark Career Trajectory. You don't get *any*where. Things do not progress. They just turn, man, they just turn and spin, like the moon does, and the electrons inside your thumb."

With one arm he draws a great circle in the air, then applies his thumb to the radio dial. "Here. Listen to this," twirling til he finds a droning academic monologue:

> The upsot is ah that there are now more dogs in the Los Angeles Metropolitan Area than there are people in the state of Wyoming. Estimates put the ah urban dog population at about three hundred fifty thousand, many of which are ah allowed to run lose. Now, if you consider that each of these animals has to ah relieve himself at least once each day, you can begin to appreciate the enormity . . .

Y.T. switches it off. "Can you believe it, man?"

Charlie looks at him grinning and starts to snicker. Y.T.'s big grin splits open. Charlie snorts, bulging with glee. Y.T. is shaking with it. Pretty soon tears are running down their faces, they are howling like chimpanzees, Charlie pounding the dashboard, Y.T. losing control of his car and so overcome he finally has to stop. He staggers out into a patch of high grass, collapses. Charlie is walking around the wagon, wagging his head, slamming his fist to palm from time to time, as chuckles explode in loud guffaws.

Charlie's Second Lesson

Steam leaking from the rusty hood distracts him. He lifts it and gingerly unscrews the radiator cap with one corner of his shirt. Y.T., recovering, joins him. While they watch the steam billow, Y.T. checks the battery water, unscrews the three lids,

74

shakes his car to watch the little cups of water flash down in there. Charlie helps him shake the car, then draws out the dip stick, holds it to the sun. ADD OIL the stick says. Charlie unscrews the oil cap. Y.T. starts twisting at his corroded battery connections. When they break off in his hands, he lifts the battery and shot-puts it at the grass. Plump is the sound it makes in soft earth. Charlie looks at Y.T. and his throat swells with comradeship. Round lights of understanding flash continuously in his head. Together they contemplate the black battery, its scarred leaden posts among the slender blades, white flowers of corrosion powder. Clunk is the sound of the round air filter which Charlie unscrews and sets rolling, when it hits the battery and settles.

Y.T. has torn his distributor loose, holds it in one upraised hand, a baby black octopus, tentacles dangling. His other hand is tugging at the sparkplug wires. Charlie takes apart the headlamps, then, with the screwdriver, crawls around the wagon flipping hubcaps loose, shallow bowls with Chevvy's squared-off wings imprinted. Giddily he sails these across the road, watches them wobble and spin in the morning light, each toss lifting him in some unaccountable way. He is spinning with the hubs, in slow motion, floating.

Together they drag the seats out, rip loose the splintered paneling, the floor mats. They open all the doors and windows, deflate the tires, jack up the rearend, and finally, with Y.T.'s spare tire between them like a fat gray hoop, head down the road toward some shade, where they stop and set the spare against an oak. They sit and contemplate the tire — bald, split, sun-cracked, caked with ancient mud — and after a while Y.T. quotes from Lao Tzu, who had this to say about rolling stock:

> Thirty spokes
> share one hub.
> Adapt the nothing therein to the purpose in hand,
> and you will have the use of the cart.

Charlie, still floating, likes the heady sound of this. It seems to answer everything. And yet it doesn't. He wonders what it means.

"What you said you were looking for, compadre. Space and time."

"I still don't get it."

"The spaces in between, Charlie. You must become an expert at using the spaces in between."

"Give me an example."

Rising slowly, with his feline liquid glance of zealous compassion, Y.T. says, "Okay. Okay."

Charlie had thought the dismantling of the station wagon would be the peak of this high, this swelling, drunken ardor Y.T.'s words have filled him with. But it's clear that Y.T. is not nearly finished, has not yet taken Charlie where he wants him to go. All of that was, you might say, a kind of foreplay. It is three hours later now, and they are climbing through thick underbrush, shortcutting the steep hairpin curves cut into this mountainside. The road is so badly rutted, being on it or off it doesn't much matter anyway. Charlie's feet burn. His neck burns from the steady sun. Sweat has soaked his shirt and his underwear. Yet he isn't weary. His excitement seems to mount with every step. He watches Y.T. a couple of yards above him, the rusty Harris Tweed ripped by dry branches, and dark behind the armpits, almost to the pockets, the lion's hair sprinkled with dust and bits of bark.

They are heading for Y.T.'s ranch, and while they climb he keeps yelling back to Charlie verses from Lao Tzu, and Navajo proverbs, and other lines about the basic rhythms of the universe, the seasons, the equinoctial tilt, moon phases, the evils of technology and our need to re-discover an organic relationship with this planet which is not only our vehicle and our spaceship but our larder and our natural home.

At one point, as they scramble for handholds in the sliding, rocky hillside, he shouts down, "Have you thought at all about the rise in masturbation, Charlie?"

"From time to time!"

"Or about how much trouble people have kicking cigarettes? Have you thought about nose picking? Biting of the fingernails? Compulsive cleaning of the ears?"

"Yes? Yes?" Charlie yells.

"Well, it's all related. Everything is related."

"How is that?"

"People are dying for things to do with their HANDS!" Y.T. screams.

Hearing this, and watching him claw his way through a thicket of manzanita, Charlie suspects, and not for the first time today, that Y.T. may be insane. He doesn't care. What he says makes so much sense, is so precisely what Charlie needs to hear at this point in his life, his meeting with Belinsky seems not coincidence but miracle. And when the gruelling slope levels out at last and they break past the final line of greasewood and sawgrass and redwood saplings and wasted trunks of longago felled and rotten eucalyptus trees and step into the clearing, Charlie is so caught up in this ascent, and so lightheaded from thirst and hunger, and yet so heedless of those pangs, he is in a state near ecstasy.

Charlie's Third Lesson

With pilgrim's eyes he gazes around at Y.T.'s ranch, the principle feature of which is dry yellow grass, about an acre of it, matted like his yellow hair and spotted here and there with high mounds of dirt. Over near some shade oaks stands a ruined barn, poles of sunlight pouring through its gaps. Beneath a tall stand of redwoods, Charlie sees a warped and soggy cabin that must have been rotting in perpetual shade for at least a century. Around it unnamable scraps of one-time farm equipment — blades, poles, buckled straps — lean and hang and burrow into the earth.

The clearing is so still, after these hours of brush crackle and sliding rock, it's like passing through transparent skin into a globe of pure silence. This only lasts a moment. A dog's snarl ruptures it. A pen made of old boards and chicken wire leans against the cabin. The dog begins to hurl itself against the wire.

Y.T. shouts, "Down, Rex!"

His voice drives the dog to greater fury. It runs frenzied circles inside the pen.

Y.T. springs up onto the nearest mound of dirt and, with arms akimbo, invites Charlie to come and gaze into the square hole it came from, about four feet wide, six feet deep, its sides dried to the polished roughness of welded iron. Charlie studies it for a while, looks up at Y.T. who says proudly, and squinting a little, like a true man of the soil, "This was the first test hole, man."

"For what?"

"For the well."

"Why did you stop digging?"

"Chick changed her mind."

"What chick?"

"Little water-witch I had up here for a few days." He winks lasciviously, then resumes the landowner's squint. "Said she could smell it a mile away."

"What happened?"

"She smelled it every place, man. I mean, look at the holes I ended up digging. The place looks like a goddam Practice Range for gophers."

Amused by this, Y.T. grins broadly, pulling at his beard, the fire in his eyes mixed, it seems to Charlie, with some desperate attempt to get another laugh bulging between them. Charlie is distracted by Rex, an enormous dog who still lunges at the chicken wire, barking ferociously. The pen posts crack each time he leaps.

Charlie waits for Y.T. to do something, but Y.T. is too busy leaping himself, down off the mound and across the clearing toward a great pile of garbage decaying in the sun — brown lettuce leaves, coffee grounds, lemon peels, fish heads, watermelon rinds. He kicks a melon rind, disturbing its fringe of buzz flies.

"Here's my compost heap, man. The law of return. Anything that comes from the soil goes back to the soil. You dig?"

Nearby a small piece of ground has been spaded. The overturned slabs are split and dried solid.

"You use the compost for gardening?" Charlie asks.

"One of these days, man, one of these days. Quick as we get the water system set up. Let me show you the house."

Charlie gets the impression that Y.T. no longer wants to look him in the eye. This last remark comes back over the tweedy shoulder, the leonine face obscured by a scarf-end tossed back, and Y.T. is pushing through thick grass as if his house contains the key, the explanation for whatever it was he brought Charlie here to witness.

They are twenty yards from it when the chicken wire pen collapses and Rex comes tearing out of there, a long black Afghan hound, shaggy, gaunt beneath his knotted coat, bounding toward Y.T. with lips drawn back over vicious yellow teeth. What looks to Charlie like a tow rope is tied around Rex's neck. Just as he springs, aiming for Y.T.'s chest and throat, the rope — connected to a tree beside the pen — goes taut. For an instant Rex hangs there, like a hooked trout clearing water, then he falls, howling and straining at the long, thick line.

With that fall something inside Charlie drops too. Watching Rex struggle in the dry grass, he is the drunken man turned sober, the lover suddenly cooled. He looks up from the dog to the man who led him to his high plateau, and sees Y.T. blinking. The flesh of Y.T.'s face is leaping wildly, as if trying to leave the bones.

Charlie can't watch it. He stares down at Rex, whining, snapping at the line. After a long time he hears Y.T. says, "I ought to shoot that sonofabitch."

"What for?"

"He killed my chickens, man. When I first moved up here I had seven chickens. Rex bit all their heads off, one by one."

"What do you feed him?"

Without answering, Y.T. moves on toward the house, driven to finish this tour of his domain. Inside they find ten pounds of Surplus dried milk, twenty of Surplus cornmeal, a sack of salt, a quart of rancid cooking oil, a braid of garlic, and two gallons of red wine. Y.T.'s face has calmed down. He stands next to his grubby sink, head half turned toward a lightless window, watching Charlie sideways, with lidded eyes, like a boy caught lying, unsure whether to fight or ask forgiveness.

Outside Rex is whimpering, gagging when he pushes at the limits of the rope. Charlie listens to this and looks at a pan of blackberry juice moldering on the stove.

"They grow wild," Y.T. says with a defensive grin. "I was going to make some jam, but I didn't have any sugar."

Spinning Things

It takes them four hours to reach a paved road. They have been hitchhiking for two more, without success, when they come upon a closed service station, standing by itself where a small mountain canyon opens onto the Coast Highway. A dim light shines in the empty office, silhouetting the pumps. Another light, fluorescent, glints off a lone red Mustang parked inside the glass-doored garage, front end pointed toward a bank of tools, a high shelf of tires wrapped in paper, and a collection of silver hubcaps hanging on the wall like trophies.

A passing motorist would see one more light, far from the other two, next to a field at the station's perimeter, where Charlie now stands dialing.

After five rings he hears the click, the operator confirming a collect call, and then Faye's thin, filtered voice saying, "Charlie? Charlie?"

"Yes. Can you come and pick me up?"

"Where are you?"

"I don't know."

"What do you mean, you don't know?"

"What does it sound like I mean?"

"Have you been drinking?"

"Yes."

A long silence. Nothing but the wire's hum, the distant bleeps.

"Can you tell me the general direction?"

"North."

"North of where?"

"San Luis Obispo. Up Highway One. It's a Standard Station. You go past Morro Bay."

"My God, Charlie! That'll take hours."

"I don't mind waiting. I want to see you. I want to talk to you."

"I want to see you too, Charlie. I mean, it's hard not knowing

79

". . . and . . ." Her voice starts to break.

"We have to have a long talk . . . about everything."

"Are you all right? Do you have a place . . . ?"

"I want you to bring some dog food, Faye."

"What?"

"Horse meat. Purina biscuits. Anything like that, as long as there's a lot."

"Listen, Charlie, I've had a hell of a rough day." Her voice breaks again.

"I know it, honey. The dog food's for a friend . . ."

"It was enough just trying to get that tire changed. And then the Triple A man turned out to be some kind of sex fiend, kept trying to get me inside the back of the camper. I mean, you can imagine how it looks, a woman off by herself, with a big double bed, and all that equipment."

Charlie can imagine it quite easly. All the way down the mountain he has been wanting Faye. He is jealous of the Triple A man. He must make love to Faye in the camper at least one time before they sell it.

"That's all going to change," he says. "Believe me, honey. We're going to start over. We're going to simplify everything. We just have to get this guy's dog fed, as a kind of a favor. I owe him a lot. And I'd like you to see this piece of land anyway . . ."

Faye is sobbing quietly now.

"He changed my life," Charlie says.

"You'd better give me the number you're calling from."

"Probably the worst fuckup I've ever known."

"What's the area code?"

"But everything he says makes sense. It's uncanny. He's almost like a brother to me now."

"The number, Charlie. Please. Just in case. And tell me what to do after Morro Bay."

He tells her, and gives her the number, kisses the receiver, and hangs up, then takes a good pull from the gallon jug they carried with them from the the ranch. Red-eyed with fatigue, he slumps against the phone and stares across the station lot at the hubcap collection inside the garage. Something about the repetition of all those circles, the way they catch the light, is hypnotic. Hard to tell at this distance whether certain hubs are fixed or spinning. Rings, bowls, blades and spokes. Charlie stares so long his eyes glaze over, doubling the images.

Seen from the road he would appear to be oddly marooned. No sign of a parked vehicle that might have brought him here. No other living creature in sight. A passing motorist, if there were one on this deserted stretch of highway at this time of night, would not see Belinsky, who lurks just beyond the office light's weak penumbra, leaning against the Men's Room door, revelling in the longest piss of his career. He would only see Charlie, dozing inside a booth whose light marks the zone between the station's asphalt and the darkness where the field begins, below the looming ridge.

I told Addie it wasn't any luck living on a road when it come by here, and she said, for the world like a woman, "Get up and move, then." But I told her it wasn't no luck in it, because the Lord put roads for travelling: why He laid them down flat on the earth. When he aims for something to be always a-moving, He makes it long ways, like a road or a horse or a wagon, but when He aims for something to stay put, He makes it up-and-down ways, like a tree or a man. And so he never aimed for folks to live on a road, because which gets there first, I says, the road or the house? Did you ever know Him to set a road down by a house? I says. No you never, I says, because it's always men can't rest till they gets the house set where everybody that passes in a wagon can spit in the doorway, keeping the folks restless and wanting to get up and go somewheres else when He aimed for them to stay put like a tree or a stand of corn. Because if He'd a aimed for man to be always a-moving and going somewheres else, wouldn't He a put him long-ways on his belly, like a snake? It stands to reason he would.

ANSE BUNDREN *in* As I Lay Dying,
by WILLIAM FAULKNER

The Odyssey of Charlie Bates

I. *1968*

PARTS ARE DROPPING onto the freeway, fuel pumps, radiators, loops of rubber tubing. A few cars are slowing down. Charlie can't take time to figure out what the trouble is. He's late for work. He only hopes that shrewd driving will get him past it and into the clear again. Like the Indianapolis winner in a racing-car movie he saw a while back, he takes the wheel firmly in both hands, widens his eyes til the sockets hurt. He swerves around a fallen transmission, cuts past the car it fell from, flicks his eyes to glimpse a stunned face, and instantly regrets the flick. The car in front of him has suddenly stopped. He brakes and spins his wheel, catches the left rear fender, bounces into another lane. Horns blare. He steps on the gas. Nothing happens. His engine is a washing machine full of clattering rocks. All around him swells a cacophony of blaring, bleating, trumpetting horns, while cars plow into one another. Greasy chunks of machinery clutter the lanes like a field of meteorites.

Brakes scream behind, and Charlie tenses. Someone slams into him. Underneath the hood he hears a wrenching clunk. He looks around for more careening cars. Most of them are idling now, or dead. Drivers in all the lanes are climbing out, kicking uncertainly at the loose parts, peering under the cars. Charlie climbs out, squats and looks at his engine block sitting on the pavement, plug wires dangling. Water pours from radiator. Like blood from an artery, gasoline spurts out mixing with the water.

He steps around to look at his rearend. The other car has crumpled it, grill to trunk, in a coupling that for an instant looks obscene. A voice behind him says, "Meetings like this sometimes seem inevitable, don't they."

The driver has joined him, a woman in her late thirties, cool, dark-haired. He's a little wary of the mystic look she gives him, as if testing which occult frequency he might be on. But she is very attractive, in her flared yellow trousers, low-cut Mexican shirt, large metal mandala pendant pressing at her cleavage.

Charlie's glad he wore his leather vest today, and the God's-eye cufflinks, and dropped his sideburns another inch. He pulls at his moustache, as if appreciating her remark. He says, "I think they call it karma."

She likes that, asks mysteriously, "What are we going to do?"

"I'm going to get my lunch and hike out through that tunnel."

She looks at the tunnel, about fifty yards away. It leads into town. Other drivers are already heading for it. She stares at the dark, round opening, smiles and says, "You are a scorpio."

"That's right," Charlie lies, locking his car.

She blushes. "Am I blushing?" she asks.

"Yes."

"It is the gonads."

"The gonads?"

"Scorpios are governed by the gonads. It's a sign of deep-seated and powerful sexuality."

They are hiking toward the tunnel now, stepping over brake shoes, around crusty axles. Charlie has to admit, it feels pretty good — the pattern broken, inevitable forces making him late for work. He is prepared to accept that this busty woman walking next to him is, as she says, among those forces beyond his control. He likes the random companion. It gives the day an edge. His thighs flex, his arms swing. As they enter the tunnel, eight lanes wide and arching grandly overhead, Charlie feels on the brink of something.

A lot of drivers are already in there, striding under murky orange lights, carrying lunches, shopping bags, attache cases, knapsacks, portable tapedecks and television sets. Most of them are complaining about the cars falling apart, asking one another what the hell is going on. Charlie doesn't participate in this. Nothing much surprises him any more. He listens to Antonia reading from the newspaper:

> SCORPIO — Now everything has to be complete and fully visible. Hidden matters are suddenly disclosed — but not all the details. You can make extraordinary progress if you move promptly.

By this time they are moving very slowly. More drivers have pushed in from behind, and an unaccountable multitude is milling as far ahead as Charlie can see. Stalled cars and trucks are scattered along the way. He figures this is what's holding things up. Soon the crowd isn't moving at all, eight lanes packed, and Charlie finds himself shoulder to shoulder, face to face with a bearded young man whose fire eyes study his, flicking anxiously. Like Charlie he wears a moustache, but it isn't trimmed. Hairs cover his entire upper lip, so that he can chew on them with ease. He seems to decide something about Charlie. Through his moustache he mutters, "Listen, man. I have a fantastic idea."

"What is it?"

"We can blow this fucking tunnel sky high."

"With what?"

"I've got the stuff. I've got it on me, man. A time bomb. All we have to do . . ."

"Hold it! A time bomb? Why do we want to do that?"

"This goddam place is gonna cave in any minute, that's why. The cars all broke down, didn't they? How long do you expect this fucking tunnel to hold together? We're already running out of air."

It could be true, Charlie thinks. Pressure is building in an odd way, as if the walls might be contracting.

Before he can reply, a blast of music startles them both. The man whirls as if caught from behind. Some kids have shoved identical reels into several tape decks, sending drums and tambourines and loud guitars pulsing from wall to wall. The song is called JUICE. A few bodies begin to jerk to the urgent lyrics:

> *It's like a wide wide flood*
> *Pouring down the bloodstream*
> *Of my mind.*

Antonia says, "You want to dance, Charlie?" With her eyes she makes this a test of his scorpiality.

For a moment he thinks about his job, or perhaps not getting there at all. A tiny pang. Gone. He looks around for the bomber. Gone too. The prospect of being blown to bits at any moment arouses him. Gazing into Antonia's gypsy eyes, their lids black-rimmed with liner, he says, "I guess so. Sure. Why not?"

There's no room to dance, but dozens of couples are trying to. Not much else to do in here right now. It's mostly heads bobbing, shoulders tossing. Charlie keeps searching for the bomber. Antonia distracts him. She is unbuttoning her Mexican shirt, long glossy fingernails unlooping each wood button while her eyes challenge him to match this stunt.

Charlie's shirt is a blue and green alligator pattern, the full sleeves sheening phosphorus. As he unbuttons it, Antonia throws hers aside. Other women begin to strip, and several young girls. Everyone glows orange in the subterranean light, pulling off the coats, the shirts and ties, and shawls, and blouses. Charlie takes off his vest, and his alligator silk. Under it he wears a candy-stripe T-shirt. He pulls this up to his armpits and rubs his belly against

Antonia's, in time to harmonicas, throbbing electric cellos.

A photograph appears in front of him, color shot of a nude couple outdoors on a blanket, with their heads between each other's legs. The woman, who's facing the camera, looks remarkably like Antonia. Leering, Charlie asks her, "What do you think of that?"

She unzips her yellow trousers, lets them drop.

Erotic photos have popped up everywhere, emerging from the shopping bags and attache cases, from the pockets of all the falling coats, books full of pictures, packets of glossy eight-by-tens. Charlie feels an arm slip around his waist. It isn't Antonia's. Someone unbuttons his jeans. He can't see who. Antonia wiggles up close, bites his moustache, while another arm pulls him gently backward into a sweaty forest of legs and bellies. Eager hands caress him, and he would like one to be Antonia's, although he really doesn't have a preference now. Through the bank of speakers booms a tape called "The Sounds of Love." Groans, and squeals of pleasure, grunts, giggles, maniacal laughter.

The tunnel has become a sargasso tangle of amber flesh. Whips have appeared, leather gloves, handcuffs. Poster-size close-ups of genitalia are being scotchtaped to the walls when Charlie spies, up above the crowd, on the roof of a rocket-shaped sportscar, the anxious bomber chewing his moustache and brandishing a .38 revolver. He's the only one in sight fully dressed — blue workshirt, navy dungarees. He's yelling something, but the tapes drown his voice. He starts shooting out the lights. Before anyone can stop him, fifty yards of tunnel turn black. Women howl. Men bellow. The tapes stop, as more shots are fired, from other pistols. On all sides, barrels flash out at the darkness. Bullets ping off metal hoods, ricochet along the walls.

Charlie drops to the floor, hands protecting his head. He finds himself on top of a naked body.

"Sorry," he murmurs.

"No need to apologize."

A woman's voice. Not Antonia's. She's lying on her side.

"Helluva situation," Charlie says.

Someone falls across his ankles. He tries to shift his weight, inadvertently strokes the woman's haunch. "Oops."

"It's all right," she whispers.

He leaves his hand there. Her lips happen to be next to his ear. He'd like to see her face, but this dark is total. There's no adjusting your eyes to it. Lights are going out much farther down, one by one, little smudges of distant orange, suddenly shattered. No one strikes matches, for fear of becoming a target. The shots get louder, not closer, but larger, as if from rifles now. All Charlie can see is the nearby screen of a small portable TV set, which silently depicts the darkness of the tunnel, broken by noiseless flashes, like tiny sparks.

"I wish we could see that set a little better," he says.

"Sometimes it's nicer when you can't."

He feels her turning under him. She too is still aroused from the dancing and the pictures and the threat of extinction. Charlie quietly mounts her. Around them other invisible couples are pairing off.

II. *The Tunnel*

Afterward he lies there listening to the crossfire, a muffled popping now, which seems to be moving down the tunnel, back the way they have come, with a strange absence of cries, just the popping, as of firecrackers going off, somewhere around a bend in the blackness. Part of the crowd begins to stir, groping to its feet, then edging along in the eerie flick of matches, a few flashlights playing on the walls and roof.

Some, reversing their tracks, turn to follow the sounds of battle. Others stay where they are, a grotto of voluptuaries writhing loosely, and watching themselves on portable screens. The rest, Charlie among them, push farther into the tunnel, grabbing what clothes they can from the shirts and trousers and dresses thrown off during the dancing. Like a man who has eaten too much, Charlie is both full and empty, drained, impatient to get away.

The woman stays close, hip to his thigh, hand clutching and re-clutching his. She whispers, "the walls look like they're getting closer together."

He has observed this too. In the irregular light these walls seem curiously active. It's unnerving. Especially now. Impending disaster no longer titillates him. All that spilled out back there in the dark. He understands the desperation in her voice. Yet in an odd way the walls are soothing. He imagines he is heading into some kind of cornucopia, away from its gape-mouthed tumble of profusion, moving toward the center, toward the source.

"Maybe it's an optical illusion," he tells her.

"I'm scared."

"Don't be. As long as we can keep walking . . ."

"The walls aren't supposed to get closer together."

"Listen, let's talk about something else. What's your name?"

"Myra."

"Mine's Charlie. I thought that was really nice back there, Myra. You know? In spite of the gunfire and everything."

"So did I, Charlie. How old are you?"

"Thirty-five."

"No kidding."

"How old are you?"

"Thirty-five."

"I'll be damned," Charlie says. "Did you go to highschool around here?"

"Yeah. I went to McKinley."

"You didn't graduate in '54, did you?"

"Yes I did. That was the year."

"Wait a minute. Wait a minute." He cranes in close to see her features. She's wearing someone's red cheerleader sweater. "You're not Myra Nordquist."

"That used to be my name. Sure."

"Wow. This is incredible."

"What do you mean?"

"I used to masturbate a lot in those days," Charlie says.

"I guess we all did."

"And for about a year there, you were the girl I would think about."

"Well then, it's almost like the old days, isn't it, Charlie."

He puts an arm round her shoulder, pleased that he has calmed her some. The coziness of this reunion lasts about thirty seconds. From behind them, what sounds like a detonation comes rumbling down the tunnel. It rumbles from a long way back there, more like the echo of a detonation, or the pre-echo of something about to explode. The pavement quivers.

"My God!" Myra cries, "What was that?"

"Probably a time bomb." He turns to squint into the flicker and gloom. Myra shrinks, pressing closer to him.

"Oh Charlie, what does it mean?"

A thick shoulder shoves at Charlie's back, pitching them forward.

"Hey, take it easy!"

"Move it, buddy," someone mutters, "move it, move it."

All around them people are muttering, shoving, urging one another to hurry. Charlie finds himself in a mob surging, then suddenly piling up where part of the tunnel is blocked.

A '52 Studebaker sedan sits sideways across the lanes.

In the crush of bodies the man behind has wedged his head between Charlie's and Myra's. "I don't know about you, buddy, but

I want to get out of here alive."

Charlie thinks "Don't we all," but doesn't say it, wouldn't encourage this man in any way. His tone is too insistent, a moralist aching to deliver a speech. The crush eases, and Charlie tries to push past the car. He can't make it, the passage is too narrow. He hears Myra saying hopefully, "Don't we all."

With fierce intimacy the man asks them, "You think it's an accident we're hearing bombs explode? You think it's an accident we've ended up like this, like a herd of cattle wandering around in the dark?"

A few inquisitive heads turn, and he raises his voice. "Well, it's not just an accident. No matter what some people may tell you, things like this just don't *happen!* This was planned to happen, it was planned from a long time back, and there are people right here in this tunnel who were in on that plan!"

A small space opens. Charlie tries to slip through it. "C'mon," he tells Myra. "Let's get going."

She grabs his hand. "No! Wait a minute!"

The man moves into the space and holds it with upraised palms. One dim orange lamp illuminates him, as if from distant torches, glinting his wolfish eyes. Charlie wants to get away from those eyes, yet he lets Myra hold him back. This man reminds Charlie of his highschool football coach, stocky, grim-jawed, his suit gray, his tie gray, his shirt white, his thin hair dark and combed straight back. He attracts in the same sort of way, with a draw both fraternal and sadistic. Little curves of white saliva coat the corners of his mouth, like the coach's always did in the locker room at halftime. "Need," the coach would growl, his voice permanently gravelled by a backhand to the throat during his last game as a semipro, "you have to NEED your opponent. You have to want him so bad you can't eat breakfast."

Charlie notices that, apart from maybe two dozen listeners, most of the crowd pays no attention at all to this man. They just inch along, waiting to get past the car. In close, people are admiring the hubcaps and door handles and other chromium features. The car is brand-new, still bears the paper license plates.

The man is shouting now. "All I want to say is this! If we expect to get out of here, then we'd better FIND OUT WHO THOSE PEOPLE ARE!"

His scattered audience, a little larger now, sends up a ragged cry. Seizing the moment he pushes past Charlie and in toward the Studebaker, throws its doors open. The driver still sits there, a slender man in bermuda shorts, drinking coffee from a thermos and listening to Frankie Laine on the radio:

A rose must remain with the sun and the rain
Or its lovely promise won't come true.
To each his own, I've found my own . . .

"Why is this guy sitting here blocking the way," the man in gray demands, "while the rest of us have to hike God knows how much farther?"

There is no explanation, and other men crowd the open door, shouting at the one inside. They wear crewcuts, short sleeve sport shirts. So does the driver, who looks up from his thermos with a cautious smile. Charlie can't help disliking him a little. The car *is* a nuisance, no matter who's responsible. And with a leader like the man in gray it's easy to turn on anyone who comes along. It happens, though, that Charlie's partial to this song on the radio. The words make a lot of sense to him, he feels a bond with the man who tuned it in. He is edging closer, drawn by both men, when Myra shoves in front of him, pulling at sleeves and collars in a frantic push toward the car.

He grabs her shoulder, handful of cheerleader's red, ribbed wool. "Myra! What are you doing?"

She jerks free, shouting back, "Things *don't* just happen, Charlie. There has to be *some* kind of explanation for all this!"

A moment later she's helping three men drag the driver out of his Studebaker, screaming obscenities, while the victim cries,

"Hey! Look out! You're spilling my coffee, for God's sake! Hey, what's going on?"

"Myra!" Charlie shouts. He lunges to get between her and the others. Someone pushes him. Someone else trips him. While he's falling, a knee catches him in the ribs. A forearm drives his neck toward the pavement. He tries crawling for the open door, but a thicket of angry feet stop him. He rolls away, and he's up on all fours scrambling for his life through churning legs, which quickly give way to the measured pace of pedestrians filing past the car.

Scuttling he swings wide around the front fender, sees a space, joins the passing stream, and soon finds himself part of a strong, almost tangible current now running in the tunnel. It flows past his calves, his wrists, his temples, an inexorable push. Or is it a pull? Hard to tell. The walls are still contracting. Yet there seems to be more space. Things are breaking loose.

Once well away from the car he glances back. Someone has turned the radio up, probably to drown the noise of struggle. From here it sounds like a muzak selection, the lulling rhythm, the styleless piano. But the way some people sway with it, scrape their feet, Charlie can't help thinking that this is the original article, and these are the days before markets and restaurants began piping in the muzak. At the edges of the throng, couples are foxtrotting, cheek to cheek, slow and easy. *Old Buttermilk Sky.*

Up close to the new Studey, car-lovers crowd the windows, peer in at plastic seat covers, aeronautic dashboard controls. Charlie can't see the driver, or Myra, or the man in gray, and he is glad it's too late to do anything about them. Or too early. He doesn't trust himself, that close to the man in gray. He's glad this current has him now. He lets it carry him.

He strides ahead. Everyone is striding, gradually picking up speed. Charlie's almost trotting when another bottleneck slows him down again.

A convoy of army trucks has stalled, ten or twelve of them, their hoods thrown up. GI's in fatigues hang over fenders trying to discover what's wrong. Heaps of ruined parts have been shoved under the running boards. The tunnel is four lanes wide here. Two MP's try to keep the crowd moving past the trucks. But a lot of people want to stop and talk to the soldiers, who sit on the roofs with carbines in their laps, or lean out the canvas-lined backends.

A drawing of Betty Grable decorates the door of one truck. She's lying on her side with legs outstretched, wearing shorts, the truck labeled *Betty's BrASS BANDITS*. Girls climb in and out over the tailgate, letting themselves be fondled and abused by the soldiers, passed from hand to hand like sandwiches. Charlie can see them in there, with sloppie-joe sweaters bunched up around their necks. Lots of hikers are stopping, just to watch, leaning back against the opposite wall as if they intend to stand there a long time.

Charlie would like to watch a while himself. He spent two years of his own in the army, right after highschool, as a Motor Pool Dispatcher, his world the color of green olives. And the canvas covering these trucks brings it all back to him, the orderly pattern of those days, his life in other hands. He often thinks of his commanding officer, a captain in the middle years who loved that color, swam in it. His face tinted olive he would chew the mushy stub of a brown cigar and tell Charlie about the great days in North Africa and Berlin.

A billyclub pokes at Charlie's ribs.

"Hey soldier."

"What?"

An MP is glaring at him. "Where's the rest of your uniform?"

93

"I'm not a soldier."

"I said where's the rest of your uniform?"

Indignant, Charlie looks down at his clothes, sees that he is wearing someone's GI fatigue jacket over the red and white candy-stripe t-shirt.

"Look, sarge, this is . . ."

The MP has him by the arm, marching him toward a jeep at the end of the convoy where two officers sit with boots up on the dashboard, sipping from a canteen they pass back and forth. One of them looks familiar.

"Caught this man out of uniform, sir."

A captain squints hard at Charlie. His face tinted olive, he wears a web-covered field helmet, speaks around the mushy stub of a brown cigar.

"What's your name, soldier?"

"I'm not a soldier."

"Goddam it, that's no answer!"

It warms Charlie to be spoken to this way again. He grumbled as much as anyone else, but he always secretly enjoyed the orders, the pointless hostility. He almost comes to attention, half hopes the mention of his name will trigger some response, some of the old gut-brothers camaraderie.

"Bates, Captain. Charlie Bates."

Behind the squinting lids, inside the harsh reply, he thinks he detects a flicker of recognition. Perhaps not.

"You realize there's a war on, Bates? We're not running a goddam fashion show for fairies here. This happens to be a combat outfit. Is that the way you want to look when the krauts are breathing down your neck? All duded up like Harry the Fairy? You look like a peppermint stick. Where's your uniform, for Christ sake?"

"I don't have one. But it's easy to explain . . ."

The MP says, "We need an extra man to help Morgan, sir."

"Right. Good idea, sergeant. Give him Harry the Fairy."

"The name is Bates, Captain. Charlie Bates?"

"See how the old peppermint stick feels after he's been down there in the gravel and grease for a while."

Morgan is a sharpshooter from Texas, assigned to guard the equipment littered beneath the trucks. On his belly, with a carbine cradled in his arms he squirms along from axle to axle, walking his elbows in front of him. Goaded by the MP's billyclub, Charlie falls in behind Morgan, cradling his own carbine, studying the tread on the soles of Morgan's combat boots. Each time they cross the opening between two trucks, Morgan pauses to stare up the skirts of the girls standing by the fenders. "Lookee there," he calls back to Charlie, "that one aint even got no pants on."

Compelled to prove his manliness, Charlie guffaws and pokes his rifle barrel in the girl's direction. Morgan shouts "Hoo-eee," then head-signals Charlie to follow him under the next truck.

For the first few yards Charlie's pleased he can keep up with the big Texan. Taking this kind of punishment should put the lid on that Harry the Fairy routine.

After three truck lengths, though, the pavement is cutting into his elbows and his knees. His neck is getting tired. He decides to reason with Morgan. "Hey, why do we have to crawl around like this? Why can't we just walk alongside the truck and crouch down once in a while, something like that?"

With a terrible scowl Morgan hisses back to him, "Whatsa matter boy? You chickenshit?"

Morgan doesn't wait for an answer. He wriggles on, and Charlie admits silently that he *is* out of shape. He remembers seeing that same look the times his old C.O. would lament the practice of bringing wives overseas, the addition of the padded backrest to the patrol jeep. "Chickenshits have got control of the world," he would cry.

That complaint had a strange appeal for Charlie. He recalls how his own capacity for abuse increased until at times it seemed a career in the army was the only sensible way to spend his life. Now, intimidated by Morgan and the MP and the captain, Charlie knows he could be crawling around under here for days. He has to get away.

Inside one of the trucks, a scratchy record starts to spin, the Vaughan Monroe band, with a militant brass section and heavy drum work, playing "The Caissons Go Rolling Along." As they squirm toward the next open space, Charlie holds back, watches Morgan lift his head, like a cobra's, transfixed. Some girls are jitterbugging with the soldiers, skirts flying, arms flailing like jump-ropes. Morgan is exclaiming, "Shee-it" and "Hot damn, thank ya ma'am!"

Picking his moment, Charlie slides out under the running board, into a flurry of booted feet and saddle oxfords. He flings himself into the dance, looks around, sees that no one has observed him. The big MP is back there letting a teenage girl handle his billy-club. Charlie tears off the fatigue jacket. Ducking he dips between the whirling dancers and the line of trucks. He's heading for the trickle of pedestrians when he hears the MP shout, "Stop that soldier!"

Charlie bolts, sprinting for the shelter of darkness up ahead.

"That homo in the striped t-shirt," the MP screams, "he's trying to desert!"

Glancing back, Charlie sees the MP raise his forty-five. On top of the trucks carbine chambers go clickety-click. Barrels nose out the windows. From the crowd of watchers along the wall someone yells, "Get that bastard!" And the nearest men rush toward him, shouting about deserters and queers. Carbine fire chases him down the tunnel. Hunched like a commando crossing a mine field, Charlie zigzags from wall to wall.

Finally the shooting stops, the shouting dies. The spectators troop back to their places, drawn by the dancing and the uniforms and the trucks themselves, where the guns are, and the music. Charlie can still hear it faintly:

> *Over hill, over dale,*
> *As we hit the dusty trail,*
> *The caissons go rolling along.*

The music fades. It's quiet, except for scuffs of feet against the pavement, the breathing. Charlie's once again among the hikers, with the spaces between them getting wider all the time. Everyone is jogging now, toward a show of light up ahead, a faint gray haze across the darkness, evidently coming from around a bend in the tunnel.

Charlie starts to run. He almost misses a cardboard sign propped against a dilapidated plywood trailer. The sign says GRIDDLE CAKES. A worn woman in a dusty dress hunkers next to a Coleman stove. Above her stands a man in overalls cleaning off a tin plate with a newspaper. He is lean, weary, slack-skinned. A tattered double mattress is tied to the roof with twine. Another one like it, limp as a handkerchief, covers a rusty Plymouth in front of the trailer. Something about the woman makes Charlie stop.

He says, "How much?"

"Five cents each," she says, without looking up. In her voice he hears the defeat of someone long past hoping anyone will buy. It touches him. He orders four. The way she stirs the batter, he knows she doesn't expect him to hang around long enough to see them poured. He hunkers next to her and sets a quarter on the apple box she uses for a table.

This lifts her spirits. She says, "How you like em?"

"However you cook em. I trust you to cook em right."

She smiles up at him then, and Charlie sees she's not as old as

96

she looked at first, around twenty-eight, twenty-seven maybe. Blond hair, searching eyes. They gaze at each other for a long moment. She tells the man, "Leon, keep an eye on these griddle cakes, will ya?"

With great effort Leon squats by the stove, his sallow face contorted. The woman beckons Charlie to follow her through the trailer door.

Dropping the latch hook into its eye, she tells Charlie to make himself at home. There's no place to sit but on a stained mattress matching the ones outside. Underwear and old hats hang from nails above it. Movie posters decorate the walls, showing Jean Harlow in a transparent gown, Jean Harlow in a satin jumpsuit, Buster Crabbe as Tarzan. Charlie looks closer at the woman who hung them there.

"You want a drink of water?" she says.

"I'd love a drink of water."

From a clear gallon jug she pours some into a metal cup, watches him drink.

"Not bad," Charlie says.

"You're the first one to stop."

"Well, it's not the best location in the world."

"That's what I told Leon. But he had his mind set."

"Can you fix anything besides griddle cakes?"

"Truth a the matter is, I did not invite you in here to talk about griddle cakes."

Charlie waits.

"Leon is . . . well . . . incapacitated. In lots of ways. And like I said, you are the first person that has bothered to stop at all."

He doesn't know what to say. She is almost past her bloom. But he imagines that in another dress, another room, she could still be a fine looking woman. The need in her eyes is so great, he is moved to take her hand. She squeezes hard. With the other she begins to open the front of her faded dress. Her breasts are full and high and, noting Charlie's stare, she explains, "Don't know whose fault it is, me or Leon's, but we aint never had no kids."

A tiny tug at her hand and she is sitting next to him. "Nor much of anything else," she adds.

Leon's battered voice comes through the door. "Fanny, these griddle cakes done all burned to hell."

"Cook up some more then. I'll be right out."

"I can't."

"Why not?"

"The burnt ones done stuck to the griddle."

"Well, scrape em offa there."

"It's smokin up so bad I can't git in close enough to see what happened."

Smoke is now seeping through cracks and into the trailer. Charlie breaks for the door, ripping the latch out as he shoulders through. He leaps into a cloud of smoke, where Leon is coughing, flailing his hat at the stove.

"Get it away from the trailer?" Charlie yells.

"What?" Leon yells back. "What? What?"

"The whole trailer's gonna catch," Charlie shouts, and he kicks the stove about ten feet back up the tunnel.

"Hey," Leon growls, "why'd ya do that?"

"Waddya mean, why did I . . ."

"Goddam it ta hell, buster! That there's a fine old stove!"

Charlie turns to Fanny, who is standing in the doorway with her dress half buttoned and looking better to him all the time. He turns to Leon, smoky-faced, who glares around, rubbing gnarled hands up and down the thighs of his ancient coveralls, uncertain what to lunge at first, Charlie, Fanny, the trailer, the stove, maybe one of these non-customers ignoring him, racing for that haze of light around the bend.

Leon doesn't frighten Charlie, he just depresses him unbearably. Too weak to fight. Too stubborn to bargain with. Glancing toward the haze Charlie says, "I guess I'd better get going."

"I'm comin with ya," Fanny says.

Without conviction he tells her, "You'd better not do that." Then a wistful smile, a regretful wave, and he turns, loping away.

To take his mind off her plight Charlie watches the haze grow brighter as he rounds this last long curve. The tunnel is still narrowing, two lanes now instead of eight, and it actually seems to be squeezing him toward the opening, as through a tube of ointment. He's almost there when Fanny catches him, barefooted, shoving back strands of hair made silky by the silver light.

"I got nothing against Leon," she explains, out of breath. "It's just that he aint *never* going nowhere else. He likes it there. And I plain don't."

He's glad she caught up in time to share this moment with him. Like a membrane dividing light and dark the opening seems to catch and hold the whiteness. Rushing through, Charlie almost feels a resisting tug. It lets them go, released, and they are shouting, spurting into sunshine.

III. *Where His Grandmother Lives*

Wondrously at ease they bound along, nearly floating. The air itself is buoyant, sweet, clean, a heady foam of invisible bubbles. Beyond the tunnel a sidewalk begins, lined with oaks. They slow down. A few cars stand at the curb, Packards, Model Ts, long and open-topped Pierce-Arrows, elegant in the brilliant afternoon. A new Stutz Bearcat waits all by itself. It is low, bright red, ready to go. Something about the Stutz Bearcat has always appealed to Charlie — the way its convertible top sits above the cab like a natty golf cap, the way you reach outside to pull the brake, and no doors to open. The ultimate roadster. Charlie unscrews the glinting radiator ornament, finger to the water line. He tries the horn, kicks a skinny tire, bends and drags his fingernail across the spokes.

"Tired of walking?" he asks Fanny.

"Not yet."

"I am."

He helps her in, cranks it started, hops into the driver's seat and sits there feeling its engine rumble. A hat hangs from one of the dashboard knobs, a green wide-brimmed felt, with a broad green band. Winking at Fanny he tries it on, cocks it. She winks back, reaches over to sharpen the angle. Charlie yearns to put this car in gear and roar away. He knows he can, the car belongs to him if he wants it. The chance of a lifetime. If only he didn't already know where it would take him. He thinks he is going to cry.

"Where we gonna go, Charlie?"

Tears start dripping down his cheeks. He is pounding on the steering wheel with open palms, muttering, "Goddam it anyway, goddam it to hell."

"What's the matter, Charlie?"

"Nothing. C'mon."

Killing the engine, he dries his eyes. They climb out and walk down the street to a bicycle shop where he rents a two-seater, purple and white, with two sets of high handlebars. •"A lost art," he tells her as they climb aboard, "four legs working together."

Other travellers from the tunnel meander down the paved boulevard in two's and three's. Charlie weaves among them, melancholy, until he hears a Dixieland band, sitting on the steps in the shade of an old boarding house, playing one of his favorite numbers. The clarinet tweedles, the banjo chunks, the trombone waddle-a-dahs. This sound revives him. They stop to listen. Still wearing his green felt hat, Charlie tips it back to take a vocal on the second chorus:

When the red red robin
Comes bob bobbin along, along,
There'll be no more sobbin
When he starts throbbin his old sweet song.

The boys in the band nod and smile, and when the song is over, they pass him a linament bottle full of homemade gin. Fanny takes a modest sip, Charlie takes a long pull. As they pedal off down the boulevard again, he feels a great deal better. Behind them the piano player taps his foot, the band swings into "That's a Plenty."

Pavement gives way to cobblestones, flat and neatly fitted. Down a broad cross-street Charlie sees a regiment of Doughboys marching rapidly at the head of a long silent parade, brown trouser legs bound above the boot, thin weapons stiff against the brown woolen shoulders. Silent billows of beige confetti spill from the rooftops, swirl around enormous brown and white flags and grotesque effigies of the Kaiser. A few tunnel people peel off and hike down that way for a closer look.

Charlie keeps pedalling until they reach a public park — thick-limbed maple trees, men in straw hats and hard collars strolling in the shade with women wearing dresses long enough to touch the grass. Lemony warmth coats this scene like a fluid, edging maple leaves with lemon light, slowing everything down. Around the perimeter cyclists cruise through honey. In the middle of the park stands a domed pavillon, and a hometown band is playing "On a Bicycle Built for Two." Charlie hums along, hears Fanny humming a nice little alto to counterpoint his baritone.

People from the tunnel are still heading down the boulevard. Charlie squints into the distance to see what lies farther on. In the next block he sees carriages, canopied buggies, a few rickety electric cars, and one steam-powered touring car with plush diamond-pleated upholstery. Then the cobblestones become a dirt road, and there are buckboards, wagons, ox-drawn carts. Way out past the edge of town, almost farther than he can see, herds of horses seem to be kicking up the dust. He doesn't want to go that far. He likes it right about here. This suits him fine. His sideburns work, and his moustache. He can trade in this felt for a derby or a straw. As for his red and white t-shirt, in this light it's not at all gaudy. He reminds himself of a brightly striped fish he once saw at the aquarium, behind thick glass, and back-lit, gliding through green-gold water.

Fanny likes it here too. She stops humming and smiles at him with sly complicity, as if they have both gotten away with something. "Ya know, Charlie, when me and Leon first started west, this is where I thought we were going."

Her smile is familiar. Seeing it he instantly recognizes the place. "Hey. This is where my grandmother lives."

"How do you know?"

"There she is," Charlie says, spotting her among the languid strollers hanging on the arm of her beau. Her waist is cinched tight, she carries a yellow fan, hair piled high on her head, and she smiles the same sly smile.

"See her? Over there?"

Fanny says, "I don't much like that dress she's got on."

"Neither do I."

It's the one thing about this scene in the park that bothers him, the long dresses, the arms hidden under taffeta. But Charlie figures everything has its price. And public practice, of course, is not necessarily the private truth. Fanny, he recalls, warmly, has a taste for the see-through lounging pajama, the Harlow look. Now that he thinks of it, she resembles Harlow some — the pentrating eyes, the short blond hair, like platinum in this luminous air. This is what attracted him in the first place, back at the trailer, that look on her face. He sees it clearly now — a cross between Jean Harlow and his grandmother. The way Fanny smiles at him he suspects she has been wearing a pair of satin pajamas all along, under the dusty shift.

His grandmother and her beau disappear into a grove of trees. Charlie turns off onto a cobbled street that borders the park, pedals past the bandstand, savoring the tuba section. Above the park the road climbs into shimmering foothills of dandelions, alfalfa meadows. He decides to ride up that way with Fanny, to stretch out a while, in the high alfalfa, behind some cottonwoods. Later on this evening, when it's cooler, they'll be back for more of this, the tuba's chesty, reassuring cadence, the glow of golden brasswork beneath the dome.

Three miles away a minor civil servant was stationed at the barricade. He saw the mayor's helicopter pass over him, decided that the freeway was now open for business, and swung the barrier out of the way. A thousand ordinary citizens, who had been detouring around the construction area for months, delightedly barreled straight ahead up the hill over the virginal concrete, throttles to the floor.

The first wave got to the dedication site four abreast, and flat out. The tail end of the parade hadn't even started, white doves were still emerging from their cages, musicians still casing their instruments, and the freeway was still filled with horse trailers; the public grandstand was still in place and blocking three of the four southbound lanes.

The front-running motorists, splendidly conditioned through years of freeway driving, reacted instantly by locking their brakes and laying down long smoking lines of rubber. The chain reaction set in through the following ranks, fenders crumpled, headlights shattered, one swerving pick-up truck took out eighty feet of the pristine fence, and the whole new highway section came to a shuddering halt.

from TAKE AN ALTERNATE ROUTE
by Paul "Panther" Pierce

Gasoline

CHARLIE BATES is speeding down the boulevard in search of gasoline. He has forgotten why he needs it. He is like the diver who has stayed below too long, kicking for the surface with that urgency near panic. The diver doesn't think about what he uses air for. He only knows his lungs cry out for lots of it and soon. So it is with Charlie and his car. The tank is almost empty. Far to the left his needle flutters over the lonesome letter E.

He knows his needle well, knows its every habit. The fluttering means he still has a gallon, maybe a gallon and a half. In times past he might have toyed with this, pushed on down the road testing how many miles he could make after the fluttering stopped and the needle flopped over playing dead. It was a little flirt with destiny he used to love. Today he can't afford it. Such pastimes depend on stations at every corner, unflagging supplies of fuel at all hours of the day and night. Some say those days and nights are gone forever. On this long boulevard, at eleven o'clock on a Thursday morning, each station he passes is closed or not pumping. It makes Charlie's eyes itch. It is like waking up in the wrong country, or on the wrong planet.

For as long as he can remember, the stations have been there, like the streets, like the sky. From the earliest days of his childhood the pumps have beckoned. Charlie can recall the time when managers gave you things, rewards for buying gasoline. Here, take this plate, they would say, take this set of dishes, this quart of Pepsi, this teddy bear. My pumps runneth over, they would say. Now the pumps are drying up. And no one can explain it. Some say the world supply of fuel is running out. They wag their heads and say your next tankful could be your last. Charlie doesn't accept this. Others say that oil profiteers, perhaps the Arabs, are to blame. He isn't sure about the Arabs. He has never met one. But as he speeds along past the laundromats and taco bars, the drive-in car washes and the drive-in banks, the muffler shops and stereo warehouses and tire outlets, and as his anxious stomach begins to flutter like the needle on his gauge, it helps to have someone specific to point a finger at. It occurs to Charlie that Arabs might be lurking behind all our addictions. Coffee. Hashish. Horse racing. Sexual excess. Gasoline.

A gauzy layer of fumes and car heat hugs the boulevard. In the near distance a yellow pole emerges from this layer, and atop the pole a large square sign says MARTY'S GAS AND GO. His heart leaps. Something tells him Marty's is open and pumping. Some vibration rises through the street fumes to charge the morning air around the sign. As he nears the corner and the station comes into view, Charlie's heart leaps again. He sees cars lined up beneath a metal canopy that shades the four pump islands. He sees the hoses that connect each car to a pump, and the drivers who stand holding hose nozzles shoved into their cars while they watch numbers change inside the little windows on the pumps. Charlie is astonished at how good this makes him feel. It is like arriving home after some exhausting journey.

A single line of waiting cars snakes outward from the pumps and down a cross street that meets the busy boulevard. From his vantage point it looks to be at most a line of thirty-five or forty. This doesn't bother him. Half an hour's wait is a small price to pay compared to lines he has heard of in other regions, or compared to no fuel at all. At this station it could be less than half an hour. They only pump gas. No one takes up precious time checking under the hood. There are no mechanics, no racks, no batteries for sale, no fan belts or road maps. Just the four islands of four pumps each, tied by computer to the tiny white office set back from the canopy, where the cashier sits watching lighted numbers come up on her console. Most days the cashier is the only one on duty. This morning a young lad has been hired to direct traffic on and off the quarter-acre asphalt lot. He stands there in his long blond hair, his jeans, his t-shirt that says TODAY IS THE FIRST DAY OF THE REST OF YOUR LIFE, and flags cars between his apple-crate barriers whenever spaces open at the pumps.

As Charlie eases past the waiting cars he removes his dark glasses so he can look directly into the eyes of his fellow motorists. He sees a sense of well-being there that seems to match his own. He thinks this is more than smug anticipation of the full tank now within reach. In the way these drivers willingly take their places, in the way they obey the blond attendant, he sees long-starved communal instincts rising to the occasion. It comes very close to patriotism, this sharing of small inconveniences to keep the larger show on the road. Charlie is touched, and reassured. An unexpected rush of comradeship makes his eyes water. His vision blurs. He has reached the next corner and swung wide, preparing to u-turn and pull up behind the final car, before he realizes the line doesn't end at the corner. It turns the corner, evidently to avoid blocking the intersection, and continues up a side street.

He sees another forty-five or fifty vehicles, maybe more. He sees campers, delivery vans. His throat goes dry. His eyes, damp moments ago, are also dry. Blinking, squinting to see how far the line extends, his eyes burn in the dry heat. The block behind was a mix of homes and small businesses, a chiropractor's bungalow, a pet hospital. This street is all residential, part of a subdivision, with small lawns, new trees spaced between interchangeable ranch-style houses. The street curves, and the line of vehicles follows the curve out of sight.

As he moves along the line Charlie sees a different look on the faces waiting. These folks don't have Marty's pumps to spur them. Trees and houses block their view of Marty's big sign catching sun-light. He sees boredom here, edging up on anger. By the time he has passed a dozen cars there is fury in the eyes of certain people who try to stare him down as he rolls by. He slides the glasses back on, just as a woman about twenty sitting in a dune buggy with an exposed engine gives him the finger for no reason. Three cars later a man in a rusting white Mustang with spoke wheels backs up a foot and rams the next car in line, a perfectly restored 1961 Cadillac hearse with paisley drapes across the rear windows. The Cadillac driver leaps out. Charlie observes these two men in his rearview, toe to toe and ready to punch.

Then he loses sight of them. He is following a long, suburban horseshoe loop. Two-thirds of the way around, after he has passed seventy-seven more vehicles, including numerous pickups, an Airstream trailer and a two-ton rent-a-truck, another street branches off, cutting deeper into the district. Again the line takes the corner, rather than clog the little intersection. This too is a curving street, and again the line curves out of sight.

The houses now are a few years older and more expensive. The trees are higher, shading the sidewalks and the larger,

107

greener lawns. This used to be a walnut grove. Full-grown walnut trees stand near each house, mingling with the acacia, the liquid amber, the date palms, and the spreading ivy. Someone has parked a roadster at the curb. It causes the line of waiting cars to bulge like a boa constrictor with undigested prey. There is a craziness about this manuever. These drivers enjoy the opportunity to create some true congestion. As he inches past, Charlie enjoys being forced to the opposite curb. He laughs to himself. It keys him to the mood along this section of the line. These people are feeling reckless. The pumps are so far away, their plans for the morning have all been scrapped.

He passes a van with bubble portholes and flames curling outward from the wheels. It is filled with highschool kids dressed for the beach. They have given up ever getting there. Giggling helplessly one girl has sprawled on a sloping lawn under one of the walnuts. Two others dance next to the van sipping German beer and inventing disco routines to the drums and bass line thundering from tapes inside.

A few cars ahead he sees a motorbike rigged like a vendor's wagon. It is parked in the street near a Plymouth Horizon, and the woman straddling the seat holds a white styrofoam cup under the spigot of a five-gallon thermos. Charlie slows down. He opens his window. The aroma of fresh coffee pours through. For hours he has smelled the rubbery false air coming through his vents, and the false leather mustiness his naugahyde upholstery gives off when the sun hits it. This coffee happens to be French roast, pungent with chicory. The aroma is an aphrodisiac, filling him with affection for whoever brewed it and thought to bring it clear out here. Nearing the motorbike all he can see are sandals, jeans, flaming copper hair. As he stops he pulls the glasses off again. He has been told dark glasses look dramatic in some sinister and compelling way. He has a hunch that this time eye contact might be more effective. He gets them off just as she turns. Her eyes gaze directly into his, blue and confident, so confident he almost looks away. He is captivated. He wants to say something memorable.

He says, "How much is the coffee?"

"Thirty cents a cup. No refills." Her voice is soft, her eyes brim with merriment.

"For French roast," he says, gaining control of himself, "that's a pretty fair price."

"I make my money on the baklava."

He glances at the wide shelf above her handle bars where the gleaming pastry squares are stacked in a white carton, their layers of translucent crust thickened with honey and grated nuts. He inhales. The holiday sweetness blending with chicory stirs all his appetites to life.

It makes him bold. He says, "What else do you do?"

"I sometimes read cars."

"You mean, professionally?"

"Your radiator, for instance, it is on the verge."

"Of what?"

"And you yourself," she says.

"On the verge?"

Her words start warm light rising through his body. Charlie lives alone at the moment. He has been married, divorced. The job he works at pays the bills, but it means less and less to him. For months now, years perhaps, something inside, elusive yet urgent, has been pushing toward the surface. He has felt himself ever nearer to some momentous threshold.

A horn beeps. He looks in the rearview. Cars and trucks are stacking up behind him. "Sonofabitch," he mutters.

"Catch you later." She seems ready to burst out laughing, not at Charlie, but at the very way life unfolds.

As he stomps the accelerator, roaring ahead, his first thought is to park quickly, walk back and find out what she meant. A glance at his gas gauge reminds him there isn't time. It feels immoral to drive away from such a woman. Yet any delay now would be too costly. A gallon left, and Marty's is the only station open. He squints hard to quell the fierce itching in his eyes. Something solid is slipping out from under him, something as firm and as fixed as the asphalt. In the old days if you were interested in a female, a car was an advantage, your strongest ally. Maybe those days too are gone, like the free dishes, and the ever pumping pumps.

For a while he holds onto the hope that this line will end within her route, so he can park and wait until the pastries and the burnished hair come putt-putting into view. Pushed onward by his dying needle and by the gas-seekers piling up behind him, he guides around the bulges where local cars are parked. With every curve in this stalled and serpentine caravan he expects to see the final car. He follows the street through a blighted tract where uprooted stumps and scaffolded foundations mix with gutted cottages half torn down. Beyond this the line slopes over

109

a rise and down into a neighborhood set apart by a broad thicket of Monterey pine. Here the landscape changes dramatically. The colors change. One side of the pine thicket is gray with construction dust, the other side intensely green. The houses are quite elegant, two stories high, colonial in style, with white porch pillars and manicured lawns. Tall redwoods rise in groves behind the houses which are like manor houses fronting large estates, all tastefully gathered around a cul-de-sac where the line finally ends.

Charlie has passed hundreds of vehicles, perhaps a thousand. He figures he has travelled at least two miles, more likely three, when he pulls up behind a boat trailer attached to the rear of an immense and glistening El Conquistador motorhome. This is not part of the neighborhood, it is the last vehicle in line. Through its blue tinted glass he can see vinyl furniture, overhead area lamps, a twelve-inch Sony. Up one side an aluminum ladder climbs to the roof where the heating and cooling vents emerge and some deck chairs are strapped. The trailer bears a twenty-two foot fiberglass launch with red padded seats and a hundred-fifty horsepower Mercury engine hanging from its stern. Together the launch and the Conquistador occupy one fifth of the available curb space. Above the simulated knotty pine door of the motor-home phosphorescent paste-on letters say NESBIT'S REVENGE.

Seconds later another car pulls in behind Charlie, then another, and another, until he can look straight across at a young fellow in a cowboy hat driving a shiny, high-wheeled Ranchero pickup with twin spots and a roll bar. He smirks, as if disdainful of Charlie's Dodge. Charlie knows it isn't disdain. He has seen this smirk on the faces of overweight women walking out of ice cream parlors with triple-deckers. It is the look Charlie wears on his own face now, the guilty smirk of the gasoline junkie who must drop everything to drive this far for a fix. Soon the cul-de-sac has filled with cars, as the line begins to loop back upon itself and down the opposite side of the curving street. The smirking, shame-faced drivers are all shutting their engines off, breaking out newspapers or switching on the top forty while they wait for the next move forward toward the distant pumps.

Into this arena of small, tentative sounds comes one large, insistent sound, a plaintive grunt, a steady grinding. He listens. It is right in front of him. He leans out the window. It comes from the El Conquistador. He hears the engine catch and start to purr, then sputter and cough, and purr a moment longer, then

110

cough again, and snort. Two smokey farts pop up between the boat trailer and the simulated knotty pine door. Then the long motorhome falls silent.

A few moments later the driver's door slowly opens. A large, heavy man climbs out, wearing a denim yachting cap, an orange jumpsuit and hiking boots. This, as Charlie rightly assumes, is Nesbit. He slams the door and stands with hands on hips frowning at the other drivers. They are all watching him. There is nothing else to do. He turns and kicks his front tire ferociously.

From the pickup the fellow in the cowboy hat lets out a rebel yell. "She's down!" he calls. "But you ain't whupped her yet!"

Nesbit kicks the tire again. He turns to his aluminum side ladder and kicks it so hard the bottom rung tears loose. Twice more he kicks it, shouting, "Goddam fucking Arabs!"

Exuberantly the cowboy yells, "You've hurt her that time!"

With a surprising burst Nesbit hurls himself against the door of his motorhome, like a lineman going at a tackling dummy. He bounces off, leaving a deep dent in the sleek beige body. Several drivers call out or honk in appreciation. Then Nesbit hunkers down next to his injured ladder and looks back at Charlie, suddenly calm, as if this has been an audition and Charlie is now supposed to make some kind of decision.

There is a delay of thirty seconds while the two men regard each other and while something inside Charlie relaxes, something he has been wrestling with since he drove away from the coffee vendor. His junkie shame dissolves. His eyes stop itching. Leaving her was not immoral, because there are two moralities, two codes in the air today, and he feels wondrously poised between them, on the verge, as she put it. Surely this must be the meaning of her prophesy. Under the Old Code you always knew what to do because in a world dripping with gasoline there was always plenty to go around. The New Code shifts from moment to moment, in a world of uncertain fuel supply. The New Code, for example, says you do not get involved with anyone whose nine-mile-per-gallon motorhome has run dry at the end of a three-mile gas line. The Old Code says any time you find yourself right behind another vehicle in trouble, no matter how many other drivers may be in the vicinity, you are the one most obliged to help. If he followed the New Code Charlie would pull around in front of this motorhome and move ahead with the line. The problem is, the line isn't moving. It hasn't moved in some time. The cars up front stand bumper to bumper with no space for squeezing in. Until someone else moves, there's nowhere to go. So for the moment the Old Code prevails, and it pleases him that he can choose to respond in the old and tested way.

He sticks his head out the window again. "What happened, pal? You run out of gas?"

Nesbit snorts a hopeless laugh. "Do dogs bark? Could Einstein count to ten?"

"You don't have a spare can?"

"Of course I have a spare can. I always carry a spare can. These days only a numbskull would drive around without a spare can." He speaks loudly and distinctly, as if from a stage. "It just doesn't happen to have anything in it. Is that a crime?"

Charlie gets out of his Dodge and looks around. He feels like stretching. He might walk over and talk to Nesbit. In a second-story window of the nearest house he sees a drape pulled back and a shadowy figure evidently watching this invasion, perhaps watching Nesbit in particular, since his equipment now blocks the driveway—a vast slab of concrete leading forty yards through shrubbery to a three-car garage with wisteria crawling across its shake roof.

Stepping toward the motorhome Charlie says, "As soon as this line starts to move, I guess we ought to at least push you ahead a few feet."

Gloomily Nesbit says, "This is not a sportscar you're looking at."

"Five or six of us could get it rolling."

This idea seems to depress Nesbit. He shoves fingers into the corners of his eyes. "Five or six of us. Jesus Christ! What a grotesque situation. I swear the goddam Arabs are going to bring us to our knees."

"You think the Arabs are behind this?"

Nesbit yanks down the chest zipper on his jumpsuit, as if the question makes him break into a sweat. He begins to fan his hairy chest with the loose orange lapels. "Do mice have legs?" he cries. "Did Benjamin Franklin wear glasses when he signed the Constitution?"

The fellow in the cowboy hat has left his pickup door open and strolled across to take a look at the dent Nesbit made. He squats and reaches out with reverence, touching the edges where hairline cracks break through the beige paint job. Instinctively Charlie hunkers next to him, so that they make a trio hunkering down out of the summer sun's fierce rays, in the shade cast by

112

the big motorhome, picking at the asphalt and discussing the mysterious forces around them, as men have done for thousands of years when thrown together by the comraderie of mutual affliction.

"I got a cousin drives a truck," the cowboy says. "He tells me oil companies are holding fuel back just to run up the price."

Charlie says, "I have heard people say the whole thing is a false problem."

"You call this a false problem?" Nesbit says. "Down at the ass-end of a deadend street? Five hours from the gas station, and no way to get there?"

"What I mean is," Charlie says, "some guy has supposedly invented a car that gets a hundred and thirty-four miles to the gallon."

Nesbit takes this personally. "I don't see you sitting in one."

The cowboy says, "Oil companies bought up all the patents."

Charlie says, "I heard the CIA bought up all the patents."

"Same difference," the cowboy says. "Those clowns in Washington all sleep with the oil execs—according to my cousin."

Charlie has a fresh idea to test out. "You ever get the feeling that we are the last generation of suckers?"

Fresh for Charlie, this is a revelation for Nesbit. He slaps his forehead with violent ecstasy. "Oh my God! Did you say suckers? That is the understatement of the century! What's your name anyhow?"

"Bates. Charlie Bates."

"Listen, Bates. Guys like you and me and the cowpuncher here, we are the greatest suckers in the history of the world. Look at this. Fifty-seven thousand dollars is what I got tied up in all this gear, and I am absolutely at the mercy of some sheik over there in Mecca or wherever the hell they live, holding on to the oil until the price runs up, then bringing the fucking profits over here to buy our own country out from under us. You want my opinion, the real truth is, it's the Arabs and the Russians who are sleeping together. They would all like nothing better than to bring us to our knees!"

Standing up he backsteps across the asphalt for a better look at what world market conditions have done to his investment. "Look at my goddam motorhome!" he shouts. "Is there any excuse for this? Look at that ladder! Look at that paint job! What do you think it's going to cost me to get this damage fixed?"

The cowboy leans in closer to the wrinkled body and grins.

"That is truly a dent and a half, old buddy."

Charlie says, "You really think it's the Arabs and the Russians together?"

Nesbit throws his hands wide and yells at the sky, "Do bears make big poopie in the woods?"

Beyond him a glint of movement catches Charlie's eye, a glint of hope. At last the line could be moving. He stands up to see past Nesbit. Grill to license plate the marooned cars still wait. This glint comes from a black Chrysler limousine squeezing through the straits where the two rows of vehicles have bottlenecked the entry to the cul-de-sac. The limousine is long and sleek, recently waxed. A moustachioed chauffeur is driving. Tinted windows make it hard to see who's in back. The windshield catches glare. The hood ornament is one foot from the Conquistador when the Chrysler stops, without a sound, and the triple-width garage door lifts open forty yards up the driveway.

Nesbit mutters, "Oh shit."

Charlie says, "We're going to have to scoot her forward somehow."

Nesbit's hands are in the air again. "One gallon will get this baby started! That's all I need!"

"Hey, calm down," Charlie says. "We have lots of time. I'll talk to the guys up ahead about making a little room."

"I'll pay five bucks for one gallon of gasoline!" Nesbit shouts. "Regular, unleaded, super, supreme—I don't give a cat's patootie!"

Before anyone can answer he rushes to the rear window of the limousine, to explain himself. "If the rest of these bench jockies would pitch in . . . !"

The door's outward swing interrupts him, and a cultivated voice saying, "May I make a suggestion?"

A small dapper man steps out. His face is swarthy, smooth, well cared for. He wears a navy blue business suit with striped silk tie, black shoes highly polished, short black hair, black moustache and trimmed black chin beard. His black eyes hold Nesbit with a brooding intensity that causes the big florid face to twitch.

Carefully he says, "You might unhook your trailer, then push the motorhome backward into the vacated space. This should at least give me room to pass."

As he speaks, Charlie notes another glint, from the same second-story window, which now stands half open, a glint from a small mirror or, it crosses Charlie's mind, a small weapon. He turns and catches what could be an answer from the house directly opposite.

To Nesbit he says, "Did you see that?"

Nesbit doesn't hear. He is transfixed by the dapper man. "Who are you?" he demands.

"I happen to live here," the man says with the proper control of his indignation.

"Are you an American?"

"I am merely suggesting a way to have my driveway cleared. Does nationality matter?"

"Does it matter?" Nesbit roars. "Is sunshine good for pimples? Could Babe Ruth hit?"

Other drivers are drifting toward the Chrysler. It's a diversion. The line still isn't moving. The dapper man, suddenly surrounded, pulls out a white handerchief and pats his shining brow. Glancing at the window Charlie sees what could be the tip of a gun barrel resting on the sill.

"Nesbit," he murmurs, nodding at the houses, "how hard is it to unhook the trailer?"

Nesbit sees the window. He grabs the small man by the coat. "What the hell is going on? What's your name?"

"For Christ sake!" Charlie says. "Let him go! Look at this!"

In a flash it has come clear to him, the way this cul-de-sac is set up, the way the road curves and the pine barrier covers sight lines. This is a little realm unto itself, a low-profile retreat where random visitors are unlikely. Seven houses face the street, and dim faces have appeared in other windows. The way the vehicles follow the concrete rim, it is beginning to resemble a wagon train pulled into a circle with the redskins threatening. Nesbit seems to see it in exactly such terms. He glares around, then takes the dapper man by the neck and pulls him against the shady side of the motorhome, one arm on the throat, the other up behind the lean back in a hammerlock.

"Nesbit!" Charlie cries. "Have you lost your mind?"

The chauffeur has scrambled out of the limo and crouches now looking for an opening. He is an enormous fellow wearing a turtleneck sweater and windbreaker. His eyes glitter. His head is shaved. He comes much closer to the kind of culprit Nesbit has imagined bringing world progress to a standstill. Instantly hate flows between them like vapor.

Nesbit says, "One more step, I'll break his arm."

The dapper man, who cannot speak, raises his free hand commanding the chauffeur to stand back.

Charlie moves in next to them. "This is insane. All we have to do is unhook your boat trailer and let these guys into their garage."

"You think it's insane? I don't think it's insane at all! This guy is from Arabia!" He relaxes his throat hold. "Am I right?"

The man coughs a raspy, "Yes."

"What the hell are you doing here?"

"I like it here, the same way you do."

"That's not the way Arabs talk."

"I did my undergraduate work at Stanford."

"Now you own all these houses, isn't that true? You own the whole damn neighborhood, including the redwood trees."

"It happens that I do represent a corporation with a large and varied portfolio."

"And all these people peeking out the doors and windows, they are all your brothers and sisters. Correct?"

"One sister, a brother-in-law, and several cousins, all of whom will be deeply affected if any harm comes to me. Moslems tend to follow up on insults to the family."

Deeply affected does not begin to describe the chauffeur, who is staring and jerking like a caged animal searching for a way through the bars, much as Nesbit looked just before he grabbed the Arab. Charlie has a feeling the chauffeur and Nesbit deserve each other. He wishes he could leave them alone to fight it out. Why can't he? What nameless loyalty holds him here? At any moment he expects rifle fire. People are standing in the doorways peering outward. In the street a dozen motorists have gathered, some siding with Nesbit, some with his victim. The cowboy calls, "Hey, let the little fella loose! Don't he have a right to drive up his own cotton pickin driveway?"

Hearing this, Charlie sees what must be done. The cowboy is still living by the Old Code. If these were the days when a homeowner could expect free access to his driveway at any hour of the day or night, Charlie himself would by now be pitching in to unhook the trailer. But faced with a motorist unhinged by the lack of gasoline, Charlie has to search the New Code for guidance.

"Nesbit," he says. "Listen to me. Suppose we could get this wise-ass Arab to siphon five gallons out of his tank into yours."

Nesbit considers this, then says, "I want the chauffeur to do it."

Charlie looks at the chauffeur, whose jaws bunch so tightly the

116

muscles are ready to explode. "If we can get the chauffeur to do it, will you turn this guy loose?"

Nesbit doesn't answer.

Charlie says, "Look. What the hell do you want?"

"For once I want some justice! Look at my ladder! Look at my paint job!"

"You want gasoline. Right?"

Nesbit starts to laugh. "Do inner tubes stink?"

Charlie gazes into the brooding eyes of the man from Arabia for some kind of confirmation. He is amazed to find that he can read these eyes. This man is a survivor. Later he may consider retribution. Right now he just wants out of Nesbit's clutches and into the safety of his compound. After a moment the man's eyes signal his chauffeur, with a nod toward the Chrysler. "Use the reserve can."

To Charlie, Nesbit says, "You told me he was going to siphon it."

"That'll take twice as long," Charlie tells him.

The chauffeur is burning. His body glows with rage. If eyes could kill, Nesbit would now be sprawled on the asphalt covered with flies and maggots. For a long moment he tries to murder Nesbit with his eyes, then he walks to the trunk, opens it and unstraps a five-gallon can with flex spout. He has to open Nesbit's gas flap. He shoves the spout in and stands next to it smoldering.

As motorists and relatives watch and listen to fuel gurgling into the thirsty Conquistador, the charged silence is punctuated by snaps and pops from the metal can. Maybe two gallons have made this passage when Nesbit, losing patience, says accusingly, as if Charlie betrayed some lifelong trust, "I want him to siphon it."

"For God's sake," Charlie says, fed up. "Gasoline is gasoline."

"I want to see this sonofabitch down on his knees in the street sucking high octane ethyl out of the backend of that Chrysler through a hose."

With a thunk and a slosh the reserve can hits the asphalt. The chauffeur is reaching inside his windbreaker, his eyes ablaze.

The man from Arabia shouts, "Raoul! Don't be an idiot!"

Raoul doesn't hear him. He draws a snub-nosed .38 from his shoulder holster and points it at Nesbit's head. "This swine does not deserve to live."

Nesbit's big ruddy face loses its color. "Hey! Hey, hold it there! Hold it a second!"

117

His head makes an excellent target since it is several inches higher than the Arab's, so much higher that he tries in vain to shield himself. As Charlie watches Nesbit squirm for cover, his distaste turns to compassion and then to lofty detachment from the entire spectacle. Time stands still. He asks himself what he is doing in this besieged cul-de-sac at eleven-thirty on a Thursday morning. And he remembers what he set out to do so many hours ago. He was planning to sell his car. Big repair bills are right around the corner. He touches his shirt pocket where the pink slip resides, with the registration. This is really the only thing that holds him, the thing that brought him here. His car. According to the Old Code you do not walk off and abandon your car in a strange neighborhood. It's too late for the Old Code, of course, and now there is no time to ponder what the New Code prescribes. Raoul is ready to pull the trigger. Part of Charlie wants him to. Nesbit deserves punishment. But not death. The instant before Raoul fires, Charlie rushes him and diverts his aim, but he can't stop the bullet, which narrowly misses the two struggling men and blasts into Charlie's radiator.

He hears the first hiss of escaping steam. Then outcries fill the arena, as motorists retreat, or push forward. People from the houses hurry down their drives. Nesbit's arms go limp. He sinks back against his motorhome, pale, stunned.

The dapper man takes command. Holding both hands high to restrain his approaching relatives, he calls, "Back! It's all right! Get back inside!"

He then steps up to Raoul and slaps his face. "You fool! This is precisely what we do not want!"

Raoul's eyes melt with shame. "He was humiliating us."

"Give me that pistol. Get into the car."

To Nesbit the man says coldly, "You have enough fuel to get started now. Please move so I can pass."

Nesbit seems paralyzed. With glazed eyes he stares at the little Arab.

Charlie says, "Do it, man, before something worse happens."

"My life," Nesbit says weakly, "you saved it."

"Never mind your life. Fire up your engine. Move it out."

He stumbles toward the cab, yanks his bent door open with a scrape, climbs in, pumping the floor pedal and pushing at his starter like a robot.

Charlie squats in front of his Dodge to observe the flow of

splashing water. He lifts the hood and sees that Raoul's bullet continued on through his engine. The cowboy is there next to him, impressed with the damage. "Whooee," he says, "that little sucker tore right through her."

The man from Arabia appears at Charlie's shoulder. "My sincere apologies. Raoul is . . . impulsive."

"I guess the police should be notified."

"The police? What could be accomplished by notifying the police?"

Charlie turns and carefully studies the profile of this man, who in turn studies with equal care the ruined engine. Beads of perspiration hang from his gleaming forehead like tiny grapes. "I was hoping," the man says, "that you and I might reach some sort of . . . an agreement."

Again warm light starts to rise through Charlie's legs and belly, flickering toward his head. It occurs to him that this fellow might be offering an extraordinary opportunity, a chance not only to stitch up all the contradictions of this trying day, but a chance for Charlie to re-open his life. According to the New Code, when fuel is five hours away and even then unreliable, a car with a bullet through the engine is a hindrance and a yoke around the neck. Yet according to the Old Code, profit is always profit. If he could liquidate with honor, might he not then have it both ways? Might he not then take the profits and position himself for some new beginning? As if in answer, riding out of his past and toward his future, the next sound he hears is the sputtering putt-putt of the blue-eyed coffee vendor.

He looks up. The sight of her copper hair nearing the cul-de-sac elates him, fills him with high purpose. For the benefit of the waiting Arab Charlie shakes his head as if grieving one last time above an open grave. Mournfully he says, "Looks like that radiator's finished, the water pump, the short block."

The cowboy chimes in. "No tellin what all else."

"That's right," Charlie says. "Who knows what else?"

"Would five hundred cover it?" the Arab asks.

"Fact is, when I left this morning what I was planning to do was sell this car. I was hoping to get thirteen maybe fourteen hundred for it."

The Arab pulls out his handkerchief and pats his forehead. "How about fifteen hundred?"

"There is also the inconvenience," Charlie says. "Wear and

tear on the nerves. Mental anguish."

The cowboy says, "I'd sooner see a horse get shot, than my pickup."

"Two thousand is as high as I can go."

An outburst around the motorbike distracts them.

"My God?" someone cries.

"What next?" groans another.

The woman has parked in the center of the cul-de-sac and seems to enjoy the commotion she has stirred.

Charlie calls out, "What's happening?"

"Some fighting, down at the pumps," a man calls back.

Her pastry shelf is empty. She has left her thermos and her wares behind, to carry this news along the line. Charlie joins the circle just as she begins.

"The story is, two mean-looking dudes in a van hopped the curb at Marty's and pulled up in front of the lead car. When people started honking, one of the guys climbed out and stood there dangling a big machete while his buddy came around to open the tank. Some fellow filling up on the next island got so mad he just turned the hose on them, sprayed them with gasoline, then whipped out his Zippo lighter and told them to get rolling. Well, they didn't wait to see if he meant it, since he could have burned them to a crisp. They jumped back into the van."

She straddles her motorbike, blue eyes darting, amused and saddened, and teasing, or testing something in the eyes of her listeners, especially Charlie's, once she notes how intently he watches her. He isn't sure she remembers him, but his sense of destiny allows him to imagine she carried the tale this far mainly for his ears.

"On their way," she continues, "the guy with the machete whacked off four of the hoses. This didn't set too well with the folks who had been waiting since breakfast. With nobody in charge but that kid directing traffic they had to take the law into their own hands. Three cars tried to cut off the van before it got back on the boulevard, so right where people drive out, there was a four car pile-up. That's why the line isn't moving and won't be for a while yet. They have to clear the area, get those thugs out of their van without setting them on fire, and of course the hose slashing means less pumps when things resume ... "

Before she has finished, drivers are striding toward the cars, to start up and move on, or check this against radio reports. Others can't decide. "I already drove thirty-five miles this morning," one fellow complains.

"Well what the hell you gonna do?" the cowboy says, hopping into his pickup. "Camp out overnight?"

"I don't have enough gas left to get anywhere," says another, slumped against his front fender.

"We're screwed."

"We're up shit creek."

"Not me," says Nesbit, whose Conquistador is rumbling again, and whose spirits have revived. "They suckered me once! They don't sucker me twice!"

With noisy bravado he guns his engine and pulls out from the curb, finally clearing the driveway so the limousine can pass.

Raoul, wearing black shades, scowls at the steering wheel in disgrace. The Arab sits behind him again, in the velvety and air-conditioned shadows, peeling hundreds from a fist-sized wad. He hands twenty bills out to Charlie and says coolly, "Pink."

"I beg your pardon?"

"The pink slip. You wouldn't happen to have it with you."

Reaching into his shirt pocket, Charlie marvels at how often things come around. As he signs away the car his training under the Old Code compels him to seal this deal with the proper words.

"She's got a good spare tire," he says bending, with a congenial grin. "Plenty of tread. And the jack should be right there under the seat. It's the original jack."

The Arab's reply is an invisible finger to the button that controls his tinted window. It rises between them. The car purrs forward, up the long driveway and under the wide garage door which silently drops shut.

Charlie looks around. All the doors in all the houses are closed again. The lawns are green and empty. The drapes are closed. Above the roofs redwoods rise toward the hot sun, the placid sky. Unfettered, released, he is ready to float beyond the tops of those redwoods. The news from the pumps has sent him halfway there. Strangely validating this moment, that news is another aphrodisiac, like the smell of chicory. He swells with affection for the messenger. When he hears her voice again, right behind him, it is part of this floating, part of a globe that is separate from the grinding starters and thundering engines and tape decks and CB radios and beeping horns. The choked exit seems miles instead of yards away.

"Need a lift?" she says.

He turns and finds her regarding him with the same directness

121

he almost couldn't handle an hour back. The difference now is two thousand dollars in his pocket and no more gas gauge to bully him. The way she looks, he wonders if she saw the money changing hands. That could be what's attracting her. Or it could be what the sudden cash called out, the self-assurance rushing through him, the bursting confidence of a man on the verge. He doesn't much care. Whatever works. She has sized up the situation, the shattered radiator. Her merry smile is breaking loose a thread of sugary crumbs at one corner of her mouth. She has been sneaking bits of pastry. He wants to laugh.

"I could use a lift," he says.

"It's snug with two. But . . . well, it's a way to get acquainted."

He climbs on behind her, pelvis to pelvis, belly to spine. He hooks heels on the metal stirrups. "How far you going?"

"Back down to the boulevard, for more supplies," she says, revving it with a twist of the hand grip. "It's lunchtime. People are going to need sandwiches, things to drink. I also read gas stations. These folks are in a terrible fix. The trouble at Marty's is just beginning."

"You're not worried about fuel?"

"I can make this run twenty times. We get a hundred miles to the gallon, up hill. Size is the problem. That seat all right? You on there? You can hold onto my waist, you know. You'd like that. I'd like it too."

His hands explore the slender waist. A coating of gristle on the lower ribs suggests regular exercise. He figures she is thirty-four and taking good care of herself. Up close to the abundant hair he sees a few white strands you'd never notice ten feet away. As she revs it again, swinging around, he says, "How do you read these things? I mean, how did you know it was going to be my radiator?"

"How do diviners know where to dig wells? It's a gift."

He lets his hands inch around toward the firm belly flesh. She doesn't mind. A little breeze blows some copper hair back. With the breeze come erotic hints of chicory, grated nuts, and with this scent comes a vision of what could happen next. In a world of uncertain fuel supply, where conditions shift from moment to moment, her enterprise is just the sort of thing one might put a little cash into. Today of course would be the day to invest. All these drivers, sooner or later they'll figure something out. Tomorrow the line could be twice this size, or it could be gone. With admiration he realizes she has known this all along. When she

122

speaks again he knows she has been reading his mind.

Over her shoulder she says, "What are you doing this afternoon?"

"You have any ideas?"

"On my way out here I was thinking about how that line has doubled back. If I had a partner, we could work both sides at once."

"I wouldn't mind giving that a try. I wouldn't mind at all."

She is puttering forward now, past the drivers still waiting. Most have held their places, figuring there is nowhere else to go. Half a dozen chose to leave, just as a string of new arrivals reached the bottleneck looking for some place to turn around. The cowboy's Ranchero was the only vehicle to get through, before Nesbit, in his rush to escape, approached the narrow exit at too sharp an angle. His bent door changed the setting on his outside mirror, so he couldn't see the trailer. It jack-knifed, and he is boxed in on three sides, bellowing and leaning on his horn, stuck like a rhino in a revolving door.

As the cowboy curves out of sight Charlie says to her, "How would you read that pickup?"

"Gunrack is the key," she says. "Armored window, armored heart."

"How about that big Conquistador with the motor launch?"

"Anal retentive. With a short fuse."

He shouts, "Nesbit!"

The big head quivers as if slapped by Charlie's voice.

"Nesbit, why don't you relax for a minute, stand back and . . . "

"Why don't I what?" Beneath the denim yachting cap his trapped eyes bulge. "Why didn't Custer wear a blindfold at the Little Big Horn? All I need is four or five feet! Get one of these other jokers to stand back! You think I can afford to burn fuel fighting my way out of this? Jesus, what a madhouse! I'm sitting on fifty-seven thousand bucks, somebody's down there cutting the pumps to pieces, I am surrounded by Arabs, and I can't even get my goddam door open!"

His voice is behind them now, and fading. She has eased between two bumpers, up onto the sidewalk. They are taking the rise that slopes toward the next neighborhood. Soon the log jam and the cul-de-sac are out of sight beyond a curve, behind the thick border of Monterey pine. It's quieter here, just the ripples of guitar or tenor sax drifting out the stranded windows. A brief flexing in her belly tells him to slide his hands a bit farther, til

123

the fingers link above her navel. In reply she presses a shoulder back, encouraging a hug. Her hair is blowing into his mouth and ears as they clear the rise and coast a while. Then the rise flattens, and they are motoring onward past the endless line of waiting cars.